D0122332

# Environmental Careers

A Practical Guide
to Opportunities
in the 90s

# Environmental Careers

## A Practical Guide to Opportunities in the 90s

**David J. Warner**

**Library of Congress Cataloging-in-Publication Data**

Warner, David J. (David James), 1953–
Environmental careers: a practical guide to opportunities in the 90s/by David J. Warner.
p. cm.
Includes bibliographical references (p. ) and index.
ISBN 0-87371-524-1
1. Environmental protection – Vocational guidance. 2. Pollution control industry – Vocational guidance. I. Title.
TD170.W37 1992
363.7'0023—dc20 91-44374
CIP



Direct all inquiries to CRC Press, Inc., 2000 Corporate Blvd., N.W., Boca Raton, Florida, 33431.

PRINTED IN THE UNITED STATES OF AMERICA
2 3 4 5 6 7 8 9 0

Printed on acid-free paper

# DEDICATION

This book is dedicated to Samantha and Grant, in hopes of a cleaner and healthier world in which they can raise their children.

# Acknowledgments

Without the encouragement and support provided to me by several friends and professional associates, I do not believe this book would have been possible. In particular, I would like to thank those individuals who donated their time and energy to review and comment on the manuscript, including Lee Carter, Elizabeth Hamilton, Linda Kissinger and Thor Strong.

Time is a precious commodity; and to write a book requires a substantial amount of dedicated time, time that is lost to a family. So, to my loving wife, Sonya, daughter Samantha, and my new son Grant (who was born during this effort), I express my gratitude and appreciation for allowing me to pursue this endeavor during an extremely busy and challenging year.

# Preface

When I was a child my parents told me, as most parents probably do, that I could be anything I wanted to be when I grew up. However, looking at the entire universe of career possibilities can be quite bewildering for an adult, let alone a child. The only professions that I was exposed to were those of my parents, our neighbors, television heroes, and of course, a long line of teachers. If children's aptitudes and interests do not lead them into one of those limited numbers of careers to which they have been exposed, they may be in for a long struggle to obtain information on alternative career opportunities. I am one who has been through that struggle.

An early interest in outdoor activities, along with the growing visibility of the environmental movement around 1970, led me to consider a career involved with managing natural resources or protecting the environment. When I started college in 1971, there were few specific environmental curricula available, other than traditional programs in geography, civil engineering, forestry, wildlife biology, and park management.

During this time, new laws were being passed by Congress to address water and air pollution problems, to protect critical land resources, and to identify and address the environmental impacts of large-scale government projects. These laws created a demand for new breeds of environmental professionals who could conduct complex scientific studies, develop and enforce regulatory standards, and understand the broader social, economic, political, and legal aspects of environmental management.

I knew that I wanted to work in the environmental field, but was uncertain as to what educational preparation was necessary for such careers. Student guidance counselors and professors had difficulty in providing me specific advice because this emerging field was evolving so quickly. I forged ahead with my environmental studies completely unsure of the nature and type of employment which would lie in my future.

As a college student, I would have greatly appreciated information on the wide variety of environmental career options available to me. Such information might have saved me from some of the long hours spent agonizing over class choices. In the past ten years, the numerous career avenues have become much more clearly defined. Many colleges and universities have responded by developing new innovative and interdisciplinary programs to better prepare students for such work. Now that this array of choices has been fairly well established, it is time to provide students, counselors, career changers, current professionals, and even retirees with an updated view of their employment options.

Looking back over my fifteen years of professional work in government and industry, I realized that I had mentally accumulated a substantial amount of information about environmental careers. I had rubbed shoulders with literally hundreds of people who were engaged in a wide range of environmental careers. My own job searches taught me about what potential employers are looking for in candidates for various types of positions. And for the past several years, I have held positions where I recruit, evaluate, interview, and hire environmental professionals. Sharing this knowledge and experience with those considering environmental work as their life vocation is my primary motivation for undertaking this task.

This book is intended to present a broad, but concise, survey of the environmental employment picture for the 1990s. I have attempted to de-mystify the environmental field and present practical and realistic career paths for students and career changers. I have assumed that people who pick up this book have already been drawn to the possibility of an environmental career. It is not my intention to elaborate extensively on the higher philosophies of environmentalism as a reason for entering the field, nor to provide exhaustive detail on individual professions. Rather, my hope for this book is that it will simply provide a realistic overview of employment opportunities and practical guidance to those wishing to sort out the environmental career puzzle.

A Final Note:

Currently, there is much public concern across our nation about American students losing interest in science and math. Perhaps by raising the level of awareness among students about the exciting and rewarding career opportunities in the environmental field, more may be drawn into these critical areas of study. The future of our nation and planet may depend on how well we as a society can attract bright and highly motivated students into the environmental professions.

**David J. Warner**
Charlotte, North Carolina

# AUTHOR

**David J. Warner** is a Senior Environmental Scientist with Delta Environmental Consultants in Charlotte, North Carolina. He has over 15 years of professional experience in natural resources and environmental management, including employment with federal, state, regional, and local government agencies, and industrial consulting firms.

Originally from Detroit, Michigan, Mr. Warner attended the University of Michigan, where he earned a Bachelor of General Studies degree in natural and social sciences, and a Master of Regional Planning degree in environmental planning. His broad experience in the environmental field includes ten years with the Michigan Department of Natural Resources administering programs under the Coastal Zone Management Act, the Clean Water Act, the Superfund program, and the Emergency Planning and Community Right-To-Know Act. In addition, as a consultant he has assisted industrial clients in complying with the Resource Conservation and Recovery Act, the Toxic Substances Control Act, the Clean Air Act, and the National Environmental Policy Act. Beyond regulatory compliance activities, Mr. Warner's areas of special interest are environmental assessments, chemical emergency response planning, risk communication, and the community relations aspects of environmental controversies.

Mr. Warner has served on the executive boards of the state chapters of both the American Planning Association and the National Association of Environmental Professionals. He has also been a frequent guest lecturer at colleges and universities, speaking on subjects ranging from environmental regulations to risk communication techniques, and more recently, on the subject of environmental careers.

# Table of Contents

CHAPTER 1

# Introduction

# Introduction

Why environmental work? If you ask this question of practicing environmental professionals, you will likely receive answers which include words such as rewarding, challenging, dynamic, enjoyable, meaningful, and even lucrative.

People who began their careers around the same time I did were largely caught up in the ecology movement of the late 1960s and early 1970s. We saw ourselves as social activists committed to making the world a better place in which to live. Times changed, however, and many college students of the 1980s and early 1990s seem to have focused their intellectual energies on preparing for careers that would enhance their economic well-being and satisfy their material desires. With the constant onslaught of Madison Avenue advertising, this trend was not surprising. Since today's students may not be comfortable with making career decisions based primarily on values and philosophy, as opposed to material gain, the question, "Why environmental work?" must be addressed.

In 15 years of talking with friends and associates who work in the environmental field, I have come to the conclusion that most of these individuals were first attracted to this type of work because of some memorable experiences or mental images they retained from their childhood or young adult life. For me, playing in the creek behind my parents' house, taking family vacations in northern Michigan, participating in Boy Scout activities, learning to sail and canoe on local waterways, fishing with my dad and brothers, and taking a summer job at a state park during college, all contributed to an interest in outdoor activities and a deep respect for the natural world.

Although a genuine interest in nature is often one of the key motivations for people doing environmental work, there are other common intellectual reasons as well, such as working towards a better society and doing something personally meaningful.

## DYNAMIC CAREERS

In practice, environmental careers are constantly changing, which attracts individuals interested in professional growth and development. A person's career journey in the environmental field may take many unanticipated turns and twists, which keeps one's work life exciting and challenging. As a demonstration of this phenomenon, this author's career path (which is probably typical of many environmental professionals) has included a number of positions in government and industry. During his undergraduate college years, he found summer employment as a seasonal park laborer and volunteer interpretive naturalist. Later, during graduate school, he worked as research assistant for a federal/state water resources management commission. Immediately after graduation, he was enlisted as a survey researcher for a professor consulting on a government project. His first entry-level professional position was as a water quality management planner with a metropolitan planning agency, and then as a coastal zone management planner for a multi-county regional development commission. Progressing to state government, the author served in various positions with functional titles, such as land use planner and zoning specialist, grant administrator, citizen participation specialist, environmental permit coordinator, environmental program manager, risk communication specialist, and public hearings officer. After 12 years of government service, a career change into industrial environmental consulting was the author's next move. As a consultant, the author has worked in several professional capacities, including chemical emergency response planning manager, environmental scientist, regulatory compliance manager, environmental trainer, environmental auditor, hazardous waste specialist, environmental impact assessment specialist, and environmental services marketing manager.

Because environmental work is meaningful, dynamic, interesting, and rewarding, few professionals leave the field to work in other areas of employment. As an example of this, although now the writer of a careers book, this author is still employed professionally in the environmental field, and has no plans to leave it.

Coincidentally, these environmental career fields which offer such meaningful work experiences are providing good to excellent employment opportunities for interested and motivated students. In discussing the demand for environmental professionals, an important distinction must be made between those careers that are natural resources management oriented and those that are environmental protection oriented.

## The Demand for Professionals in Environmental Protection

Career opportunities in the environmental protection and management fields are expanding rapidly, driven by increased government regulation and a movement by business and industry to reduce environmental liabilities associated with their operations.

According to a paper presented at the American Public Health Association conference in 1987, a shortage of professionals to solve environmental problems was recognized as far back as the 1940s in many areas of the country. At an environmental engineering education conference in 1960, it was estimated that only 6,000 environmental engineers were available to fill a demand for 12,000. The same organization 20 years later, in 1980, estimated that only 10,000 environmental engineers were available when 16,000 were needed, a shortage of about 38%. In addressing the specific need for hazardous waste management professionals, the U.S. Congressional Office of Technology Assessment commissioned a report in 1985 which estimated that over 22,000 professionals would be needed from 1990 to 1995 to work solely on hazardous waste cleanup, and that the existing talent pool was inadequate to meet future needs. This projection was made before the Superfund program was accelerated in 1986, and does not factor in the need for professionals to perform audits, monitoring, and waste reduction activities. The 1987 conference paper concluded that the number of graduates with bachelor's and master's degrees which include hazardous waste training, will probably meet only 5 to 10% of the total demand in the next five years, and even less in following years.

In a special section on environmental career opportunities appearing in the November 26, 1990 issue of *Engineering News Record*, human resources directors from several of the nation's top environmental consulting and engineering firms reported a shortage of scientists and engineers (i.e., an extremely favorable marketplace), especially for those with five to ten years of experience. Similar sentiments have been echoed in a consulting firm's magazine where it states that consulting firms cannot locate and hire enough experienced professionals to keep up with demand, and that the situation appears to be getting worse.

A recent article by John Arundel with the New York Times News Service appearing in the *Chicago Tribune*, reports that thousands of new environmental jobs in recycling, waste disposal, and pollution control are opening up across the country, and that new college graduates are finding that, contrary to a traditional myth, they can perform meaningful environmental work and make a good living as well.

Analysts in the construction-engineering industry have acknowledged that the environmental field will be lucrative and growing for the foreseeable future, and that the market for environmental engineers, hydrogeologists, and ground water modelers, as well as those with broader skills in regulatory affairs, marketing, and finance, will be strong.

*Career Futures* magazine rates the demand for engineers as exceptionally strong, with environmental engineers leading the way. In addition, the journal suggests that practically any career with the adjective "environmental" placed in front of it will offer excellent opportunities throughout the coming decade.

One aspect of this significant shortage of environmental engineers, is that professionals with other scientific backgrounds (e.g., biology, chemistry, environmental studies, etc.) have had, and still do have, excellent employment opportunities. This trend will likely continue until adequate numbers of environmental engineers are available in the marketplace.

Tom Keinath, head of the environmental systems engineering department at Clemson University, stated in a recent newspaper article that graduates from his program can virtually pick their location of employment, their employer, and their desired salary level.

It is clear that employment opportunities in environmental protection and management, and in environmental health and safety, will be lucrative well into the 21st century.

## The Demand for Professionals in Natural Resources Management

Natural resources management careers are holding steady in general, with some specific disciplines experiencing minor growth. However, with the current tight government budgets and spending patterns due to the 1990-1991 recession, most openings are appearing through retirements, the attrition of people from government into the private sector, and individuals leaving the work force. The best opportunities are offered to those who specialize in environmental impact analysis within their discipline (forestry, fisheries, outdoor recreation, etc.). Also, with increasing public and governmental interest in protecting wetlands from destruction, particularly coastal wetlands and estuaries which support commercial fisheries, wetlands ecologists will have an improving employment picture throughout the 1990s.

There will certainly be good employment opportunities for interested, motivated students who are dedicated to a particular resource management field. However, they will need to be more persistent and patient than their counterparts studying in the environmental protection arena. The primary reason for the tight marketplace is that the majority of natural resource management positions are with federal, state, and, increasingly, local government agencies, where most programs are no longer experiencing significant growth and professionals tend to stay for long periods of time with a single employer.

A recent survey of federal and state resource management agencies by the National Wildlife Federation confirms the less than promising outlook for jobs with the federal government. The U.S. Fish and Wildlife Service, the National Park Service, and the Bureau of Land Management (all within the Department of the Interior), and the U.S. Soil Conservation Service (in the Department of Agriculture), report very limited employment opportunities, as the number of positions remains constant. The U.S. Forest Service, also in the Department of Agriculture, reports a slightly improving employment outlook in the early 1990s, as it expects to hire an additional 600 resource management professionals per year to supplement its existing force of 31,000. However, competition for these federal positions is stiff. This survey also suggests that because there are currently fewer students enrolled in wildlife management programs, the percentage of job candidates finding jobs will increase. For fisheries specialists, the Forest Service and Bureau of Land Management will be expanding their fishery programs with newly available federal funding (this is an example of how government positions are subject to the shifting nature of government budgets). Also, aquaculture positions in the private sector hold increasing opportunities for fish biologists. In forestry, the newest entry-level employment opportunities will probably be provided by industry. Urban forestry and other forestry related professions (such as forestry-related sales positions involving equipment, chemicals, and computer software) as well as providing consulting services to private landowners will provide many new opportunities.

## OPPORTUNITIES FOR WOMEN

Historically, only a small number of women have been attracted to environmental career positions, particularly those in the science and engineering areas. As such, women were grossly underrepresented in the ranks of environmental professionals during the 1960s and 1970s. This situation is beginning to change as modest, but increasing, numbers of women are entering environmental fields.

Many women have become involved in environmental work through grassroots activist organizations focused on specific or localized environmental issues. Others have entered the field as their professions (e.g., teaching, industrial hygiene, public health, and journalism) became increasingly involved in environmental subjects and issues. Female liberal arts graduates have also been successful in finding work in the environmental arena, largely through state and local government agencies and nonprofit organizations.

Today, growing numbers of women are intentionally targeting their education towards environmental careers. It is more common today than in years past to see women who are entering the work force with degrees in environmental engineering, toxicology, hydrogeology, and other technical subjects. They are finding that the market for well-qualified women environmental professionals is wide open with excellent employment opportunities.

## OPPORTUNITIES FOR MINORITIES

The employment opportunities for minorities in the environmental field are exceptional. However, colleges and universities are having a difficult time attracting minority students to technical environmental programs of study. Many universities have implemented special recruiting and scholarship programs for these individuals. Still, the pool of available minority graduates in the environmental disciplines is extremely small.

Given that minority groups are a significant and growing percentage of the population of the U.S., and that the consequences of environmental degradation are often first felt in minority communities, they are inadequately represented on the professional environmental staffs of government, industry, and the nonprofit sector. Several major corporations and government agencies have preferential hiring programs for qualified individuals from underrepresented groups. Still, these organizations report that they are unable to find qualified minority job candidates, that intensive recruiting efforts often fail, that candidates do not always understand the career potential of positions, and that retention of minority employees is a problem. In light of this situation, well-qualified minority job candidates will find that they are in high demand in all employment sectors.

The Black community is beginning to see opportunities in the environmental field. In their article, "Boom Careers for the '90s", the editors of *Black Enterprize* magazine identified several categories of "hot" careers, including: environment (law, waste management, environmental science, and public relations); consulting (especially engineers and those with technical backgrounds); engineering (includes environmental, along with civil, chemical, and other disciplines); and science/research (especially ecologists, geneticists, and biochemists). The employment prospects in the environmental field are strong for everyone who is qualified and motivated, but opportunities are even more outstanding for minorities.

## HOW MUCH EDUCATION?

Environmental work is a field which offers excellent employment opportunities for qualified individuals at virtually all educational attainment levels. The field truly does offer something for everyone. Most professional positions require a minimum of a bachelor's degree, and many employers prefer advanced graduate degrees. However, there are also substantial employment opportunities for field specialists, laboratory assistants, and technicians. Some of these positions require an associate's degree, while others do not. Often, however, some type of technical training or certification is required.

Although job opportunities for technicians are very good, those earning at least a bachelor's degree, will have the best choices for professional career advancement and the most flexibility in moving into new and interesting types of work. Some career paths will necessitate obtaining a graduate degree, either as a minimum job requirement or as a credential to compete effectively for openings. Before embarking on a specific educational program, students need to assess their interest in and aptitude for various types of environmental work. What is it that makes a certain type of environmental work attractive to an individual? Once this is determined, students need to understand all of the educational requirements for a position. The following chapters will provide a good start for this effort, as those careers in which advanced college level training is required are noted in the discussions of specific career opportunities.

Aside from formal educational training, students must be aware that the education of those working in the environmental field is a never ending process. With the fast pace of change in technologies, regulations, and scientific knowledge, environmental professionals must continually stay abreast of new developments by attending seminars, reading journals, associating with other professionals, and maintaining certifications in specific technical areas. These activities should be a consistent part of the working life of every environmental professional.

## TECHNICIAN OR GENERALIST?

One question often asked by individuals considering environmental careers is "Should I focus on a technical specialty, or concentrate more on being a versatile generalist?" The answer is, confusingly, both. Technical competence of entry-level job candidates is one of the first items evaluated by prospective employers. Students interested in environmental work, therefore, need to choose and develop competence in a technical specialty. In addition, one of the environmental professional's most valuable virtues, in the eyes of an employer, is his or her flexibility and adaptability to new problems and projects. This personal ability is enhanced by a well-rounded, broad-based environmental education, with at least some exposure to all the various aspects of environmental work (i.e., scientific, engineering, legal, economic, political, social, etc.). Such a background allows students to begin to see the "big picture" beyond the technical solutions to environmental problems. As one progresses through a chosen environmental career path, it may be seen that the generalist type educational background will become more relevant to management level decision-making. Striving first to become competent in a chosen technical area, and then to broaden and round-out one's college course work is, therefore, a sound educational strategy for an aspiring environmental professional.

## SCIENCE AND ENGINEERING: THE CORNERSTONES FOR ENVIRONMENTAL WORK

Environmental problems are often technically complex, involving the principles of ecology, biology, chemistry, physics, geology, mathematics, and engineering. Therefore, a sound educational background in the various science and engineering disciplines is necessary for environmental specialists to fully understand and solve or manage such problems. Developing management and communication skills is also critical to the success of these professionals; but the first educational priority for those aspiring to enter the environmental field should be courses in science, math, and engineering. By understanding the complex technical aspects of environmental problems, students can more fully comprehend the legal, economic, political, and social aspects of environmental work as well.

Preparation for an environmental career should begin in high school. Students at this level should take all science and mathematics classes available to them. Seniors that have opportunities to take college level science and mathematics courses should do so. The advantage of this is that these students will meet the prerequisites for the more advanced and interesting environmental courses as underclassmen, and also have the opportunity to take more elective classes in associated fields or in skill-development areas (such as public speaking, computer science, etc.). This early preparation in science and mathematics will allow students to get the most out of their college educations and provide them with a solid foundation to enter the environmental work force.

## REFERENCES

National Wildlife Federation, *A Survey of Compensation in the Fields of Fish and Wildlife Management*, 25th Edition, National Wildlife Federation, Washington, D.C., 1989, p. 1-4.

Busch, P. L. and Anderson, W. C., Education of Hazardous Waste Professionals, presented at the 116th Annual Meeting of the American Public Health Association, Boston, MA, November 15, 1987, p. 2-5.

Engineering News-Record, Special Section: Environmental Career Opportunities, *Engineering News-Record*, November 26, 1990, McGraw-Hill Inc., New York, NY, p. 4-36.

Rich, L. A., The Missing Professionals, *Resources*, The Environmental Resources Management Group, Exton, PA, vol. 11, no. 3, June 1989.

Arundel, J.(New York Times News Service), Environmental field is green with growth in employment, *Chicago Tribune*, Sunday January 6, 1991, Section 20, p. 29.

Schriener, J., Firms call for good workers but many get busy signal, *Engineering News-Record*, October 18, 1990, McGraw-Hill Inc., New York, NY, p. 30-34.

Schlosberg, J., Job Opportunity Index (Outlook Section), *Career Futures*, vol. 2, no. 2, Fall/Winter 1990, Career Information Services, Inc., Stamford, CT, p. 5-8.

Kelly, P., College seniors find job pickings slim, *The Charlotte Observer*, Saturday May 4, 1991, p. 6A.

Stover, G. W., Minorities and Environmental Career Opportunities, *Becoming An Environmental Professional 1990*, The CEIP Fund Inc., Boston, MA, p. 108.

Boom Careers for the '90s, *Black Enterprize*, vol. 21, no. 7, February 1991, Earl G. Graves Publishing Company, Inc., New York, NY, p. 55-78.

CHAPTER 2

# Trends and Prospects for the 1990s

# Trends and Prospects for the 1990s

The environmental marketplace in the 1990s, and its associated employment opportunities, will be driven by a number of factors; they include increased public awareness, new regulations, the recognition of environmental liabilities, new markets, and public demands for a better quality of life.

The resurgence of public awareness and interest in environmental quality issues, which coincided with the 20th anniversary of the Earth Day celebration in April of 1990, has pumped new energy into the environmental movement. This rising level of public interest has, for the first time, pushed environmental issues into the national political arena. Political candidates for national and state offices are including environmental topics, such as acid rain, global warming, hazardous and radioactive waste disposal, wetlands protection, and offshore oil drilling, in their campaign strategies.

Concurrently, the news media are covering environmental issues on a more frequent and expanded basis. One effect of this increased news coverage is that government environmental program managers are finding it easier to obtain financial resources from Congress and state legislatures, which, in turn, should result in the improved administration and enforcement of environmental laws and regulations. Municipalities and industries (i.e., the regulated community) will respond to these "beefed up" enforcement efforts by improving their own environmental management programs.

Another major factor driving environmental activities in all sectors, is the expansion and tightening of environmental laws and regulations (especially as they relate to controlling the release of toxic substances into the environment). The early 1990s will see the reauthorization of the Comprehensive Environmental Response, Compensation, and Liability Act (i.e., Superfund), the Resource Conservation and Recovery Act, and the Clean Water Act, as well as new regulations promulgated under the Clean Air Act Amendments of 1990. To carry out these new mandates, federal, state, and local agencies will need to increase their manpower and technical capabilities in the areas of soil and ground water remediation, waste minimization and treatment, and air pollution control. Likewise, to achieve and maintain compliance with new requirements, industries will need to employ, or retain the services of, additional environmental scientists, engineers, and regulatory specialists. Environmental interest and "watchdog" groups will also need more people to oversee this increased level of regulatory activity.

With the advent of the Community Right-To-Know Act of 1986 (incorporated as Title III of the Superfund Amendments and Reauthorization Act, or SARA), businesses must annually: (1) inventory and report on the amounts of hazardous chemicals that are used or stored on their properties, and (2) estimate and report the amount of toxic chemicals they release into the environment. This law has forced businesses to throw their doors open and come forth with information that formerly was not readily available to the public. Concerned citizens and environmental interest groups now have new information that they can bring to the attention of the news media about companies that may not be good environmental neighbors. This has prompted many companies to move quickly to "clean up their act" and "get their houses in order".

The issue of environmental liability is increasingly becoming a driving force for environmentally responsible behavior by businesses, governments, and individuals. Transactions involving the purchase or sale of commercial real estate are increasingly including an assessment of environmental liabilities associated with specific parcels of property. Companies that handle, transport, manufacture, store, treat, or dispose of hazardous substances are implementing more sophisticated materials management procedures, employee training, and emergency spill response plans to minimize risk to employees and the community. In-house regulatory audits of industrial facilities are being more frequently used as an "early warning system" to identify deficiencies in waste management and pollution control procedures and facilities. Such deficiencies can then be corrected before regulators issue violation notices and fines. It is not only the threat of government enforcement, but also the potential for law suits filed by employees or citizens (and the bad publicity that follows), that motivates companies to address and minimize their environmental liabilities.

The recent discovery and verification of large-scale environmental contamination associated with many U.S. Department of Energy (DOE) and U.S. Department of Defense (DOD) installations has opened up numerous new opportunities in the marketplace for environmental specialists. The development and production of nuclear weapons at facilities now owned and managed by the DOE, was carried out since the 1940s under strict national defense security measures. The widespread environmental impacts associated with these operations have recently been acknowledged by the DOE and an aggressive cleanup program has begun. This program, which among other things, involves the cleanup of hazardous and radioactive wastes, will take over a decade to complete and will cost billions of dollars. At another set of federal facilities, those owned and operated by the DOD, the cleanup of environmental contamination associated with the operation of the Air Force, Army, Navy, Marine, and Coast Guard installations is receiving increased attention and funding by Congress. This work will involve the cleanup of many common contaminants, such as solvents, metals, and fuels, as well as those associated with chemical warfare (e.g., defoliants, like Agent Orange, and nerve gas). An entire new generation of well-trained environmental scientists and engineers will be needed by government agencies and contractors to see these programs through to completion.

If the U.S. adopts and implements a comprehensive energy policy during the 1990s, a policy which this country sorely needs, it could have a significant impact on the employment marketplace for environmental professionals. The identification and development of environmentally compatible technologies, such as solar, wind, geothermal, and alternative fuels (e.g., methane, natural gas, alcohols), will require tremendous effort and the dedication of thousands of scientists and engineers. Finding ways to safely dispose of radioactive wastes generated by nuclear power plants will likely also be part of this effort. Because coal will still be the major fuel for electricity production in this decade, controlling further the air emissions and other environmental impacts from these facilities, as well as increasing production efficiencies, would also be addressed in a national energy policy. Increasing the fuel efficiency of automobiles would be another important policy objective. Although these activities will take place even in the absence of a national energy policy, an official commitment to such a policy would certainly be felt within the economy. The likelihood of this happening during the 1990s will depend on both political factors and international events (i.e., another energy crisis). Other factors which have and will continue to influence the environmental marketplace will be mentioned elsewhere in this book (e.g., local governments attempting to find alternatives to landfilling by recycling or by building waste-to-energy plants, and the opening up of international markets for environmental work).

On the natural resource management side, several issues will drive demand for competent environmental professionals. Urban development pressures continue to threaten sensitive environmental areas, such as wetlands, estuaries, sand dunes, deserts, and mountain habitats. Ecologists and planners are needed to identify and evaluate these critical areas and provide for their protection, or to minimize the adverse impacts of development. The scarcity of natural areas, parks, and recreation areas within large and growing cities is diminishing the perceived quality of life for urban residents. Creating new opportunities for natural and recreational experiences within urban settings will continue to challenge landscape architects, land use planners, and outdoor recreation specialists in the coming decade. With continued population growth, the increasing pressures on fish, wildlife, and forest resources will need to be addressed in order to protect the integrity and vitality of these natural resources. Natural resource biologists and managers are needed to identify and protect fish spawning areas, wildlife habitats, wilderness areas, endangered plant and animal species, and other tasks involved with striking a balance between public use and resource protection.

The point of this discussion is that there is a tremendous amount of environmental work currently taking place in the nation and around the world. Although the environmental marketplace may experience short-term ups and downs with the overall economy, the long-term prospects for environmental work point only to strong growth over at least the next ten years. This translates directly into excellent employment opportunities for environmental professionals in all disciplines.

CHAPTER 3

# Careers in Environmental Protection

# Careers in Environmental Protection

The career opportunities discussed in this chapter involve those professions dedicated to the protection of the environment from air, water, and land pollution. Most of these careers relate to the assessment of environmental quality, the evaluation of environmental impacts, the enforcement of governmental regulations, and the technologies to prevent pollution and to clean up polluted resources.

Most environmental protection work is driven by laws, regulations, lawsuits, and efforts of industry or government to reduce or eliminate environmental liabilities associated with a particular site. The major federal environmental statutes that govern nearly all environmental work include:

- Clean Air Act
- Comprehensive Environmental Response, Compensation, and Liability Act (CERCLA or Superfund)
- Superfund Amendments and Reauthorization Act (including Title III, the Emergency Planning and Community Right-To-Know Act)
- Clean Water Act
- Federal Insecticide, Fungicide, and Rodenticide Act (FIFRA)
- Hazardous Materials Transportation Act
- National Environmental Policy Act
- Occupational Safety and Health Act
- Resource Conservation and Recovery Act (RCRA)
- Safe Drinking Water Act
- Surface Mining Control and Reclamation Act
- Toxic Substances Control Act (TSCA)
- Coastal Zone Management Act
- Marine Protection, Research, and Sanctuaries Act (ocean dumping act)

Evolving from this long list of environmental laws is a growing national demand for professional scientists and engineers in both government and industry. This chapter will survey the many promising career opportunities in environmental protection work.

# ENVIRONMENTAL ENGINEER

## Description of the Work

Many different types of professional engineers, including environmental, chemical, mechanical, civil, and geotechnical, may become involved in environmental projects during their careers. However, because environmental engineers work exclusively on environmentally related work, this section will focus on their activities.

Environmental engineers apply the theories and principles of science and mathematics to the solving of technical environmental problems. In the process of developing these solutions, engineers continually invent new technologies that are more effective or economical than earlier approaches. They work to reduce the release of pollutants to the environment, to prevent pollution from harming human health and the environment, to clean up contaminated areas, and more recently, to minimize the generation of waste pollutants at their source.

The control of air pollution is one key area of focus for environmental engineers. In finding ways to eliminate certain pollutants from production processes, environmental engineers may use existing technologies and equipment such as bag houses (to capture dust and particulates), wet or dry scrubbers, or electrostatic precipitators. Or, faced with increasingly stringent air pollution limits, they may experiment with new technologies involving thermal methods or recovery systems.

The treatment of municipal sewage and industrial wastewater is another area of concern. Before these wastes can be discharged into a stream or river, it is required that they be treated. Based on the type of constituents in the wastewater, environmental engineers design treatment systems to meet the legal discharge standards in the most economical way possible. These treatment systems may involve a combination of chemical, mechanical, and physical processes to clean the water. Sanitary engineering, the term commonly applied to this area, was the predecessor to the emerging field of environmental engineering.

With the passage of the first Clean Water Act in the 1970s, the treatment of sanitary sewage by every city and town was made a top priority, and a large sum of federal government money was made available to build sewage treatment plants. As communities continue to grow, and existing communities need sewage treatment maintenance and upgrading, the need for environmental engineers in this field should grow slightly. However, federal government funding for this program has been severely reduced; states and municipalities are developing alternative funding mechanisms. This situation may temporarily delay some projects, but eventually the amount of work in this area will increase due to a built up backlog of need.

Industries also are dealing with new challenges. New testing methods and equipment are able to detect pollutants at lower and lower levels. With these changes in technology, new regulations are developed, particularly in the area of toxic chemicals. As a result, environmental engineers are being asked to find new ways to treat industrial wastewater to reduce the content of toxic substances. For example, the pulp and paper industry is struggling with ways to reduce dioxin, a by-product of a chlorine bleaching process, in its wastewater.

Managing the growing mountains of municipal and industrial solid waste presents another challenge for environmental engineers. The traditional and most common way of disposing of solid waste is in landfills. Historically, many of these landfills were simply holes in the ground (i.e., dumps). In many areas, as rainwater seeped through these landfills, pollutants were leached into the soil and ground water. Today, regulations commonly require that several environmental safeguards be designed into new landfills; these include such things as impermeable liners, leachate collection systems, and ground water monitoring wells. Many environmental engineers participate in the siting and design of solid waste landfills. With the availability of land for waste disposal becoming scarce, resource recovery (including recycling) processes and methods are being developed. Also, waste-to-energy facilities, where solid waste is used as fuel in an incinerator which creates heat to drive electrical generating turbines, are gaining in popularity and sophistication. Solid waste management and resource recovery are areas in which increasing numbers of environmental engineers are specializing.

In the area of hazardous waste management, engineers are challenged to develop new ways to treat hazardous waste to render it less hazardous or nonhazardous. In fact, recent regulations under

RCRA require hazardous wastes to be treated to certain levels, or by certain methods, before it can be placed in a hazardous waste land disposal facility. Because the thousands of regulated chemicals are so diverse, different types of treatment methods are needed. Engineers have developed hazardous waste treatment technologies which involve incineration, solidification, stabilization, encapsulation, and other types of physical and chemical treatments. Many of these technologies have been developed by engineers working on the cleanup of contaminated hazardous waste sites. As more is learned about the toxic affects of hazardous chemicals, treatment regulations are expected to become more stringent. Because this area of environmental engineering is growing quickly, and is projected to continue growing, hazardous waste management is becoming another popular specialty for engineers.

Aside from the applications of technologies to solve problems, environmental engineers also perform standard types of engineering tasks: cost estimating, project scheduling, specification development, consulting with other engineers and scientists on technical matters, construction oversight, computer-assisted design, quality assurance reviews, project management, and report writing.

## Educational Preparation

High school students interested in engineering can prepare for college by taking as many science and mathematics courses as are available. At the college level the engineering curriculum is structured with core and elective courses. Good elective courses for those studying environmental engineering include: environmental policy and law, economics, ecology, English, and public speaking.

The standard requirement for entry-level engineering positions is a bachelor of science degree from an accredited university engineering program. Science and mathematics majors may also occasionally qualify for certain types of engineering positions. Many universities now have specialty concentrations in environmental engineering, while others house environmental engineering classes within their civil engineering departments.

In addition to having an engineering degree, engineers whose work may affect life, health, or property, or who offer their services directly to the public, are required to be registered in the state where they are practicing. State registration requirements usually include: (1) an engineering degree from a university engineering program accredited by the Accreditation Board for Engineering and Technology; (2) four years of relevant work experience, often under the supervision of a registered engineer; and (3) passing a state engineering registration examination.

## Potential Employers

**Federal government** — The majority of environmental engineers employed by the federal government work for the Environmental Protection Agency, although growing numbers are being hired by the Departments of Energy and Defense.

**State government** — States carry out many of the federally mandated environmental programs, and employ substantial numbers of environmental engineers.

**Local government** — The larger city and metropolitan governments are often involved in assisting states with environmental management and regulatory programs in highly urbanized areas. As part of these programs, environmental engineers are employed by local health, public works, and water and sewer departments.

**Consulting firms** — Environmental engineers are in great demand by engineering and environmental consulting firms to assist industry and local governments with air and water pollution control, and solid and hazardous waste management.

**Industry** — Nearly every type of manufacturing industry has environmental management responsibilities under current laws and regulations. Most medium and large size manufacturers employ environmental engineers to work at both plant and corporate levels.

**Other employers** — Environmental interest groups, research organizations, and trade associations may also employ environmental engineers to conduct research, follow regulatory trends, and provide technical advice on pollution control problems.

## For More Information

American Academy of Environmental Engineers
130 Holiday Court
Suite 100
Annapolis, MD 21401
(301) 266-3311

American Society of Civil Engineers
345 East 47th Street
New York, NY 10017-2398
(212) 705-7667

National Society of Professional Engineers
1420 King Street
Alexandria, VA 22314-2715
(703) 684-2800

Water Pollution Control Federation
601 Wythe Street
Alexandria, VA 22314-1994
(703) 684-2400

American Society of Mechanical Engineers
345 East 47th Street
New York, NY 10017-2392
(212) 705-7722

American Institute of Chemical Engineers
345 East 47th Street
New York, NY 10017
(212) 705-7338

Air and Waste Management Association
Box 2861
Pittsburgh, PA 15230
(412) 232-3444

## REFERENCE

*Occupational Outlook Handbook*, 1990-1991 Edition, Engineering, Scientific, and Related Occupations, U.S. Department of Labor, Bureau of Labor Statistics, Bulletin 2350-3, p. 3-5.

# ENVIRONMENTAL GEOLOGIST/HYDROGEOLOGIST

## Description of the Work

Geologists study the physical aspects of the earth and its environment. Geologists working in the environmental field generally are involved in three major types of activity: (1) determining the suitability of land for commercial or industrial development; (2) identifying the geological and hydrogeological impacts of proposed projects; and (3) analyzing the characteristics and extent of ground water and soil contamination at a hazardous waste site, and recommending and coordinating cleanup actions.

In determining the suitability of a site for a specific type of development, geologists collect and analyze information concerning bedrock depth and composition; the characteristics, quantity, and dynamics of the ground water; seismic stability and earthquake potential; and the adequacy of the soil structure to support proposed construction activities.

When identifying the environmental impacts of a proposed project, geologists determine if and how soil and ground water resources may be degraded and what negative impacts would occur for plants, animals, and humans. They also make recommendations on ways to minimize adverse impacts on these critical resources.

When investigating contaminated sites, hydrogeologists (those geologists who specialize in ground water geology) determine where and how to take ground water and soil samples, and evaluate the results of the laboratory analyses of the samples. They also test ground water wells to determine flow pressure for pumping and treating contaminated ground water; take soil borings to verify the structure and depths of soil layers; determine the structural geology underlying a site and how it may affect the movement of contaminants in the ground water (for example, fractured bedrock, impermeable rock, sandstone, etc.); and develop and refine computer models to predict the extent of contamination, the rate of movement of ground water, and the concentration of contaminants at a downstream well site. Some use technologies such as ground-penetrating radar, electromagnetic devices, or other means to locate buried drums, tanks, and other sources of contamination. The removal and disposal of underground storage tanks (a common source of ground water and soil contamination) is supervised by these geological scientists.

## Educational Preparation

High school students interested in geology and hydrogeology should take classes in chemistry, mathematics, and biology. At the college level, the geology curriculum will include both core and elective classes. Good course electives for environmental geologists would be hydrology, soil science, ecology, organic and inorganic chemistry, computer science, and environmental engineering.

A bachelor of science degree in geology is generally the minimum requirement for entry-level geologist positions. However, geology technicians, with less than a bachelor's degree, may find positions involved with field activities and data collection. There is a growing trend for environmental geologists and hydrogeologists to earn master's and doctoral degrees, particularly among those employed as consultants and researchers.

## Potential Employers

**Federal government** — The majority of environmental geologists and hydrogeologists employed by the federal government work for the Environmental Protection Agency and the U.S. Geological Survey, although growing numbers are being hired by the Departments of Energy and Defense, as they begin site remediation work.

**State government** — States assist in carrying out many of the federally mandated environmental programs, including Superfund, and employ environmental geologists and hydrogeologists to conduct initial site investigations.

**Consulting firms** — Qualified environmental geologists and hydrogeologists are in great demand by engineering and environmental consulting firms to assist with the investigation and cleanup of contaminated sites.

**Industry** — Large manufacturing industries, particularly the chemical industry, which own multiple plants and facilities, occasionally employ environmental geologists to coordinate and oversee the investigation and cleanup of contaminated sites.

**Other employers** — Environmental interest groups and research organizations may also employ environmental geologists to conduct research, follow regulatory trends, and provide technical advice on soil and ground water protection and cleanup problems.

## For More Information

American Geological Institute
4220 King Street
Alexandria, VA 22302-1507
(703) 379-2480

Geological Society of America
3300 Penrose Place
P.O. Box 9140
Boulder, CO 80301
(303) 447-2020

National Water Well Association
6375 Riverside Drive
Dublin, OH 43017
(614) 761-1711

Association of Ground Water Scientists and Engineers
6375 Riverside Drive
Dublin, OH 43017
(614) 761-1711

## REGULATORY COMPLIANCE SPECIALIST AND ENVIRONMENTAL SCIENTIST

This is a relatively new career path necessitated by the passage and implementation of numerous federal and state environmental laws and regulations.

### Description of the Work

Regulatory compliance specialists are concerned with the understanding and application of the vast array of federal, state, and local environmental laws and regulations. Common activities for these specialists include: researching and interpreting complex regulations to determine their applicability to a specific set of circumstances or projects; documenting the status of compliance for a particular facility; conducting background technical studies; reviewing new environmental pollution control technologies for air, water, soil, or hazardous wastes that can meet legal requirements; providing guidance on appropriate regulatory procedures for proposed projects; comparing air or water quality monitoring data with permit requirements or standards; tracking the status of proposed legislation; and preparing and presenting training sessions to update the regulated community on changes in the law.

In governmental positions, regulatory compliance specialists focus on assuring that both businesses and municipalities are in full compliance with the environmental regulations administered by their agencies. Often, these professionals concentrate on a single regulatory program, such as water quality or solid waste disposal. Their daily activities might include: conducting facility inspections (e.g., a manufacturing plant, a city sanitary wastewater treatment plant, a hazardous waste incinerator, etc.); issuing notices of violation; negotiating a compliance plan and schedule for a facility to meet certain requirements; serving as an expert witness in a government law suit; providing advice to the regulated community to assist them in meeting their regulatory obligations; evaluating a remedial action plan designed to clean up a contaminated site; determining priorities for enforcement action; commenting on proposed regulatory changes; and providing certification to a local government or industry that has met regulatory requirements.

Regulatory compliance specialists working for industry, either as staff or as a consultant, play a slightly different role than their governmental counterparts. Their primary focus is to assure that their company is in full compliance with all environmental regulations affecting its facilities. The status of compliance is usually determined through an internal regulatory audit of a facility. Through such an audit, the regulatory compliance specialist can identify potential violations of air quality, water quality, toxic chemical, or waste management regulations. A plan is generally developed to address each potential violation. The regulatory specialist often works with high level company managers to obtain funding for implementing the plan. Another frequent task is preparing complex permit applications for the discharge of certain air or water pollutants, or for the treatment, storage, or disposal of hazardous wastes. Regulatory compliance specialists working in industry are also often involved in negotiations with regulatory agencies concerning compliance matters.

"Environmental scientist" is a title commonly used in engineering and consulting firms for those non-engineer, non-geologist scientific generalists who coordinate the preparation of environmental impact statements and environmental assessments, conduct real estate environmental audits, perform special research and feasibility studies, conduct risk assessments, perform regulatory compliance tasks, and carry out a wide variety of related interdisciplinary tasks. Very similar in practice and job description to what the author has termed the regulatory compliance specialist, environmental scientists may perform the identical tasks and functions. It might be expected, though, that the environmental scientist have a strong scientific background, in addition to a knowledge of regulations. However, for many employers these two position titles are interchangeable, and the only difference is semantics.

## Educational Preparation

Regulatory compliance specialists and environmental scientists might come from diverse educational backgrounds. Environmental engineers, urban and regional planners, and biologists might all have the necessary background to become regulatory compliance specialists or environmental scientists. Ideally, individuals desiring this type of work should have a sound scientific foundation from which to understand the reasoning and intent of environmental laws and regulations. Beyond science courses, additional classes should be taken in environmental law, political science, and economics. Such college classes might be taken as electives or as core curriculum courses in engineering, the natural sciences, planning, or a related field of study.

A bachelor's degree is the minimum requirement for a career as a regulatory compliance specialist or environmental scientist. Often, professionals in this field also have graduate level degrees in fields such as environmental law, environmental policy and management, or environmental science.

### Potential Employers

**Federal and state government** — Regulatory compliance specialists and environmental scientists are employed in both federal and state government. These careers will be discussed further in the section on environmental program specialists.

**Industry** — Large manufacturing industries often employ staffs, which include regulatory compliance specialists, to assist their plants and facilities with meeting their regulatory obligations.

**Consulting firms** — Engineering and environmental consulting firms commonly have job positions for environmental scientists and regulatory compliance specialists. These professionals are employed to team with other specialists, such as engineers and hydrogeologists, to solve problems for governmental and industrial clients.

### For More Information

National Association of Environmental Professionals
P.O. Box 15210
Alexandria, VA 22309-0210
(703) 660-2364

## ENVIRONMENTAL ENFORCEMENT SPECIALIST

### Description of the Work

Environmental enforcement specialists are employed largely by government agencies to enforce environmental regulations and bring violators to justice. Unlike regulatory compliance specialists, who work with the regulated community to achieve compliance with the law, environmental enforcement specialists deal with violators who are chronic, blatant, or intentional offenders, and are potentially dangerous to people or the environment.

These professionals undertake a wide variety of activities to build legal cases (primarily criminal and some civil) against environmental law-breakers. Some of their usual duties are responding to citizen complaints; conducting surveillance of suspected polluters (for example, a hazardous waste hauler illegally dumping his load into a stream); collecting samples of air, water, or soil; inspecting environmental crime scenes; gathering evidence; identifying toxic and hazardous substances; taking photographs as evidence; issuing warrants to violators; consulting and coordinating with environmental scientists and engineers in building a case; interpreting aerial photographs of criminal activity; building and maintaining case files; working with lawyers to prosecute violators; going to court to present case findings; and coordinating field activities with state and local law enforcement agencies.

## Educational Preparation

Environmental enforcement specialists can come from a variety of educational backgrounds, although a bachelor's degree is the commonly accepted minimum requirement. Law enforcement, the environmental sciences, and environmental engineering are the backgrounds of most enforcement specialists. Some even have a law degree. Because this type of work is highly specialized, the government agencies who hire environmental enforcement specialists generally provide continuing in-house training.

## Potential Employers

**Federal government** — The Environmental Protection Agency is the primary employer of environmental enforcement specialists in the federal government. These specialists may also provide coordination and assistance to other federal agencies, such as the Federal Bureau of Investigation, the U.S. Coast Guard, the Food and Drug Administration, the State Department, etc.

**State government** — Environmental enforcement specialists are employed within each state's environmental protection agency. In their work, they coordinate efforts with state and local law enforcement agencies.

### For More Information

Contact Federal and State environmental agencies for job descriptions and employment requirements. For EPA, contact:

U.S. Environmental Protection Agency
Office of Human Resources Management
401 M Street, S.W.
Washington, D.C. 20460
(202) 382-4600

## AIR QUALITY SCIENTIST

### Description of the Work

Air quality scientists (or environmental meteorologists) working in the environmental field focus on understanding the condition and dynamics of the earth's atmosphere and how air pollutants behave after they are released. Applying their knowledge of the physical and chemical properties of the atmosphere, along with information on ranges of air pressure, humidity, temperature, wind velocity, and prevailing wind direction for a specific community, they work to predict how pollutants will disperse and at what concentrations.

Using mathematical relationships between atmospheric conditions and the characteristics of particular air pollutants, these scientists develop sophisticated computer models to run alternative pollution scenarios (for example, adjusting the model to account for various rates of emissions and different smoke stack heights). Understanding the dispersion pattern of pollutants from a smoke stack, an accidental tank leak, or industrial explosion is important in protecting the community from toxic chemicals which may affect human health.

Air quality scientists also perform other tasks, including: analyzing ambient (pre-existing) air quality for a site; conducting smoke stack sampling; planning, setting up, and operating various types of air sampling equipment; measuring the amount of pollutants in the air; analyzing laboratory results of air sample testing; estimating the emissions of toxic compounds from industrial processes for right-to-know disclosure requirements; preparing air quality permit applications with data generated through computer-simulated dispersion models; developing or refining computer models to predict the migration of air contaminant plumes; evaluating the performance of air pollution control equipment; and conducting special large-scale research studies on issues such as global warming, ozone depletion, and acid rain.

## Educational Preparation

High school students interested in air quality science should take courses in chemistry and mathematics. At the college level, courses should include physical geography, meteorology, organic and inorganic chemistry, computer science, hydrology, physics, and environmental engineering.

A bachelor's degree in environmental science or meteorology is generally the minimum requirement for entry-level positions, but a graduate degree significantly increases employment opportunities, especially with environmental consulting firms, corporate environmental affairs staffs, and research organizations.

## Potential Employers

**Federal government** — The majority of air quality scientists employed by the federal government work for the Environmental Protection Agency, although some are also employed by agencies such as the National Oceanic and Atmospheric Administration.

**State government** — Most states assist in carrying out the federal mandates of the Clean Air Act, as well as administering their own state air quality programs. Air quality scientists are employed by states to implement and coordinate these efforts.

**Local government** — Larger city and metropolitan governments are often involved in assisting states with implementing state air quality programs in highly urbanized areas. As part of these efforts, air quality scientists are employed by local health departments and local air pollution control authorities.

**Consulting firms** — Air quality scientists are in great demand by engineering and environmental consulting firms to assist industry and local governments with air pollution assessments, measurements, modeling, permitting, and emissions control.

**Industry** — Most large manufacturing industries are regulated under federal and state air quality laws and regulations. These companies employ air quality scientists to work at both plant and corporate levels.

**Other employers** — Environmental interest groups, research organizations, and trade associations may also employ air quality scientists to conduct research, follow regulatory trends, and provide technical advice on pollution control problems.

### For More Information

Air and Waste Management Association
Box 2861
Pittsburgh, PA 15230
(412) 232-3444

## ENVIRONMENTAL POLICY ANALYST

### Description of the Work

Policy analysts working in the environmental protection arena perform a wide variety of functions related to the development or modification of government policies, laws, and regulations. These environmental professionals use political and legal skills, along with a technical understanding of their area of specialty (i.e., air quality, surface water quality, ground water quality, solid and hazardous waste management, environmental health and safety, etc.), to establish or influence public policy.

The duties of the policy analyst include: evaluating proposed legislation; estimating budgets for environmental protection and management programs; determining the impact of certain regulations on various affected parties; conducting policy research; building coalitions to support proposed laws; conducting economic impact analyses of government subsidy programs; strategic planning for policy development related to controversial environmental protection issues;

drafting or editing proposed policy, legislation, or regulations; preparing position papers for presentation at public hearings or legislative committee meetings; and estimating required staffing levels for proposed government programs. Additional duties may include: determining popular opinion on specific issues through survey data; soliciting input on proposed policies; facilitating consensus of opinion in committee situations; consulting individually with various interested parties; developing and evaluating alternative policy scenarios; predicting political positions by various parties on proposed government or legislative actions; analyzing regional differences in program impact; identifying critical geographic area positions on issues; and developing and maintaining liaisons with federal, state, and local agencies.

Economists specializing in environmental protection issues, also considered to be environmental policy analysts, focus on the economic aspects of pollution control and related programs. The economists' duties might involve: conducting cost/benefit analysis on proposed government or industry programs; developing sophisticated economic impact assessment procedures for certain environmental management actions; projecting the cost to business and industry of proposed regulations; developing economic incentives for waste minimization and pollution control; or, estimating the public value of specific pollution control activities.

## Educational Preparation

High school students interested in environmental policy analysis should take classes in science, social studies, and civics. At the college level, any of a number of degrees may qualify a student to become an environmental policy analyst, such as environmental science, political science, public administration, and urban and regional planning. In any case, course work should provide the student with a broad background, and include classes in science (i.e., ecology, biology, chemistry, geology), social science (i.e., psychology, sociology, political science, environmental policy and law, economics), public administration, analytical methods (such as survey research), and communication skills (such as writing and public speaking). In addition, developing a specialty area of interest (e.g., solid and hazardous waste management, air quality, surface and ground water quality) will prove useful for aspiring policy analysts.

A bachelor's degree is the minimum entry-level educational requirement for policy analysts. Many analysts continue their education at the graduate level, pursuing master's degrees in public administration, law, political science, economics, and environmental planning and management.

## Potential Employers

**Federal government** — Nearly every federal government agency employs limited numbers of policy analysts to develop, update, refine, and change policies in the agency's area of authority. In the environmental arena, the Environmental Protection Agency and certain congressional offices are the primary employers.

**State government** — State environmental protection agencies and legislative offices are the key employers of environmental policy analysts.

**Industry** — Large industrial companies and trade associations employ policy analysts to monitor the activities of regulatory agencies, to analyze the impacts of proposed legislation or regulations, to suggest changes to existing policies, and to propose new policies.

**Research institutions** — Policy-oriented research institutions, whether affiliated with universities, industry, or environmental interest groups, employ policy analysts to evaluate existing government policies and programs, monitor the implementation of government policies, and recommend new approaches to solving problems.

## For More Information

National Association of Environmental Professionals
P.O. Box 15210
Alexandria, VA 22309-0210
(703) 660-2364

# COMMUNITY RELATIONS SPECIALIST

## Description of the Work

Public controversy commonly arises when a hazardous waste cleanup project (under CERCLA/Superfund program, RCRA, or state equivalent programs) begins, or when a new site for a hazardous waste treatment, storage, or disposal facility, or any industrial plant which might cause pollution, is being proposed. When government agencies or industry neglect to inform the public (especially the affected community) about proposed courses of action, the citizen outrage that often results can cause costly delays and a poor relationship with the community. This situation is often referred to as the "NIMBY syndrome", meaning "not in my back yard!". Such situations create the need for community relations specialists that understand the technical aspects of controversial environmental issues. In fact, the current federal Superfund law mandates that an active community relations program be an integral part of cleanup plans for abandoned hazardous waste sites.

The primary concern of community relations specialists working in the environmental arena is to facilitate effective two-way communication between project decision-makers and technical experts, and the local neighbors and citizens affected by a project. Included in the activities that these professionals perform are the following: preparing a community relations strategy for a controversial project with specified goals and objectives; developing a detailed community relations plan complete with a time schedule and cost estimate; establishing and staffing citizen participation committees; interpreting and, in some cases, simplifying technical information for understanding by the lay citizen; organizing educational programs which relate to the scientific or technological aspects of a project; preparing and issuing press releases and newsletters; conducting opinion surveys in a community; researching the answers to questions posed by citizens and the news media; assisting in negotiations to resolve areas of stalemate between decision-makers and the community; preparing and implementing a press relations strategy; briefing decision-makers on issues of citizen concern; and recommending policies and procedures for incorporating citizen concerns into project activities.

## Educational Preparation

A bachelor's degree in a field related to community relations is generally the minimum educational requirement for community relations specialists. A graduate level degree is a distinct advantage for an entry-level position. Holding one degree in a communications field and another in a technical environmental field would provide the job candidate with a solid educational foundation to perform this type of sensitive community relations work. Probably more important than the name of the degree, is the combination of social sciences courses (including political science, sociology, psychology, economics, and communications) with environmentally related science and engineering courses.

## Potential Employers

**Federal government** — Community relations specialists are employed by the federal government at the Environmental Protection Agency, Department of Energy, and other agencies that get involved in environmental controversies. Most are employed in the Superfund program during the investigation and cleanup of contaminated sites.

**State government** — Most states assist in carrying out the federal mandates of the Superfund, as well as administering their own site cleanup programs. Community relations specialists are employed by states to implement and coordinate communication at these sites, as well as to assist in other controversial actions, such as issuing an air toxics permit to an industrial facility.

**Consulting firms** — Community relations specialists are occasionally employed by engineering and environmental consulting firms to assist industry and local governments with communications during controversial environmental projects.

**Industry** — Many large manufacturing industries employ community relations specialists to help establish and maintain a favorable public image in regards to protecting the public and the environment.

### For More Information

National Association of Professional Environmental Communicators
P.O. Box 06 8352
Chicago, IL 60606-8352
(312) 321-3336

National Association of Environmental Professionals
P.O. Box 15210
Alexandria, VA 22309-0210
(703) 660-2364

## LABORATORY SCIENTIST

### Description of the Work

Laboratory scientists working in the environmental field are involved in the technical analysis of hazardous, toxic, and conventional pollutants.

Using federal Environmental Protection Agency-approved analytical methods, laboratory scientists conduct various tests on air, soil, water, industrial waste, and municipal sewage plant sludges to identify hazardous constituents and measure concentrations. The results of these analyses are used to determine whether a substance will be classified as a hazardous waste or toxic chemical under regulatory definitions.

Environmental chemists test newly formulated chemicals to determine their hazardous characteristics and toxic properties in order to meet federal government requirements for the registration of chemicals. Another important task for chemists is to conduct treatability studies for hazardous waste and to develop bench-scale and pilot-scale chemical treatment methods for such substances. Chemists also prepare chemical information sheets called Material Safety Data Sheets (MSDS) to inform workers of chemical hazards in the workplace and to meet the requirements of the Hazard Communication Standard under OSHA.

Microbiologists working in the environmental field conduct studies on the impact of pollutants on microorganisms. Another area of research for microbiologists is on the utilization of microorganisms to consume the hazardous constituents of various types of waste materials.

This process, termed "biological treatment", is beginning to be used to treat contaminated soils and ground water, both on the ground surface in digesters and in the ground ("in situ" treatment).

## Educational Preparation

High school students interested in environmental laboratory science as a career should take chemistry, biology, and mathematics classes in preparation for college. At the college level, students should pursue degrees in the sciences, particularly chemistry and physics. Elective courses in hydrogeology, air quality science, and environmental engineering will round out the environmental laboratory scientist's background.

A bachelor's degree in science is usually the minimum entry-level requirement for laboratory scientists, although there are opportunities for laboratory technicians and sampling technicians that do not require a four-year degree. Laboratory researchers and managers generally hold advanced graduate degrees.

## Potential Employers

**Federal government** — Environmental laboratory scientists employed by the federal government work for the Environmental Protection Agency and other agencies involved in environmental research and investigations, such as the Agency for Toxic Substances and Disease Registry, the Centers for Disease Control, and the National Institutes of Health (all authorized under the Department of Health and Human Services).

**State agencies** — Most states employ environmental laboratory scientists to analyze samples as part of environmental site investigations and to conduct research. State laboratories are usually housed within the department of environmental protection or public health.

**Independent laboratories** — Many environmental laboratory scientists are employed by independent, EPA-certified laboratories. These laboratories are used extensively by government, industry, and consulting firms to analyze contaminants in air, water, and waste materials.

**Industrial laboratories** — Large manufacturing operations, particularly in the chemical industry, have on-site analytical laboratories. Although traditionally used for quality control purposes, many are expanding their roles to include analysis of environmental contaminants. Increasing numbers of environmental laboratory scientists are being employed in this capacity.

### For More Information

Contact individual government, industry, and academic employers for job descriptions and employment requirements. Other sources of information include:

Air and Waste Management Association
Box 2861
Pittsburgh, PA 15230
(412) 232-3444

Association of Ground Water Scientists and Engineers
6375 Riverside Drive
Dublin, OH 43017
(614) 761-1711

## ENVIRONMENTAL ENTREPRENEUR

### Description of the Work

Several businesses have been born as a result of the passage of environmental laws and public pressure to control the pollution of air, land, and water resources. Examples of environmental types of businesses include: manufacturers of pollution control equipment; environmental consulting firms that conduct site investigations, perform feasibility studies, assist with regulatory compliance activities, and otherwise provide engineering and scientific services to government and industry; architecture/engineering companies that design and build entire pollution control systems and facilities (i.e., wastewater treatment plants, air scrubber systems, hazardous waste incinerators, solid waste landfills, etc.); hazardous waste collection, transport, storage, treatment, and disposal companies; and firms that manufacture personal protective clothing and equipment for those working on hazardous waste sites.

Environmental businessmen and businesswomen tend to be experienced environmental professionals with specific areas of expertise. They generally draw upon their knowledge and experience to identify needs in the marketplace that can be met with a particular product or service. Some are the owners/operators of their own companies, some go into business with partners, and some specialize in marketing activities. Often, environmental businessmen/women perform a balance of administrative business tasks and technical environmental work.

## Educational Preparation

High school students interested in becoming "environmental businessmen or businesswomen" should first develop a strong background in science and math. At the college level, a minimum of a bachelor's degree in a technical environmental discipline (an excellent one would be environmental engineering), with additional elective courses in business management, law, and accounting would provide a solid background for an environmental entrepreneur. A master's degree in business administration with a bachelor's degree in environmental engineering, hydrogeology, or air quality science would be ideal.

## Potential Employers

By definition, entrepreneurs work for themselves. However, interested individuals would be well advised to first obtain working experience in government or industry, or in an environmental business enterprize. After gaining some valuable working experience, and when the timing and opportunity is right, entrepreneurs can start businesses in the following areas: environmental consulting, pollution control equipment manufacturer, hazardous waste handling and disposal, recycling, monitoring, and laboratory services and equipment manufacture, employee training, computer software, and other areas supported by the marketplace.

### For More Information

Contact owners and managers of environmental types of businesses. Also, read environmental professional and trade journals to identify potential environmental businesses.

## INDUSTRIAL ENVIRONMENTAL MANAGER

### Description of the Work

Industrial environmental managers are responsible for assuring that industrial facilities are in full compliance with federal, state, and local regulatory requirements. These managers may be employed at the plant level or at the corporate level. To effectively perform their duties, environmental managers need to understand the dynamics of the industrial processes that create air emissions, wastewater, solid waste, and hazardous waste.

Industrial environmental managers are called upon to conduct a wide variety of tasks, including the following: inspecting industrial equipment for chemical leaks or for malfunctions that could create unauthorized emissions to the environment; conducting an internal audit of a manufacturing plant to identify regulatory compliance concerns; preparing cost estimates to repair or upgrade machinery; coordinating the sampling and analysis of air emissions, wastewater flows, and hazardous wastes generated at a plant; overseeing the work of contractors and consultants that perform nonroutine projects, such as the removal of underground storage tanks and the cleanup of contaminated soil and ground water; conducting in-plant training of hourly employees on environmental safety and on proper hazardous waste handling practices; preparing an emergency response plan to handle accidental chemical spills; assuring that air pollution control, wastewater treatment, and solid and hazardous waste management systems are all functioning efficiently and effectively; supervision of a limited number of technical staff persons; filling out manifests for the transportation of hazardous waste to an off-site disposal facility; negotiating with government regulatory agencies about performance standards for air quality and water quality discharge permits; and preparing quarterly and annual reports for regulatory agencies on the amount of toxic substances released to the environment (such as under the federal Community Right-To-Know Act, the Clean Water Act, and the Clean Air Act).

## Educational Preparation

Industrial environmental managers may come from a variety of educational backgrounds, although the majority currently employed are probably engineers (not necessarily environmental engineers). A bachelor of science degree in a science or engineering field is generally the minimum requirement for these positions. In some smaller companies, the duties that would normally be assigned to an environmental manager are given to a human resources department employee or an industrial health and safety manager.

Beyond an educational background in science and engineering subjects, students desiring to become industrial environmental managers should take college courses in environmental policy and environmental law, if possible, while still in school.

## Potential Employers

**Corporate environmental staffs** — Most of the larger industrial corporations in the United States employ a staff of managers dedicated to environmental affairs. Usually these managers specialize in one aspect of environmental management (e.g., hazardous waste management, air pollution control, or site remediation). These executives develop company policies and provide guidance, and sometimes issue mandates to company-owned facilities. Also, they are often responsible for approving expenditures for environmental projects, both corporate-wide and at the plant level. Some companies recognize that environmental matters are important to the health of the business and establish a vice-president of environmental affairs position.

**Industrial plant staffs** — At large industrial plants, environmental managers often supervise a limited number of technical specialists who respond to chemical spills, operate wastewater treatment facilities, maintain pollution control equipment, label hazardous waste containers, oversee the unloading of chemicals into a plant, and a myriad of other related activities. At small industrial plants, environmental managers may have to perform all environmental functions by themselves. In addition, many of them have responsibilities for occupational health and safety programs.

### For More Information

Air and Waste Management Association
Box 2861
Pittsburgh, PA 15230
(412) 232-3444

National Association of Environmental Professionals
P.O. Box 15210
Alexandria, VA 22309-0210
(703) 660-2364

Other good sources are the environmental committees of industry trade associations.

## ENVIRONMENTAL PROGRAM SPECIALIST

This category of environmental careers is largely defined by the major federal environmental statutes, identified in the beginning of this chapter, that drive much of the environmental work in the United States. Rather than a specific professional title, such as environmental engineer or hydrogeologist, these careers involve developing a thorough knowledge of a specialized program area. Currently, it is not generally required that a program specialist have a specific educational background, as long as candidates have some minimum level of scientific environmental training. Any and all types of environmental professionals may become program specialists, depending on the needs and organizational structure of an employer.

The foundation of the program specialist concept is derived from the federal and state agencies designated to carry out specific environmental laws and regulations. The concept has now spread into industry, where many large corporations have environmental program specialist-type positions. Because the potential employers of environmental program specialists are the same for each type of specialist discussed here, they are presented at the end of this section.

## PROGRAM SPECIALIST: HAZARDOUS WASTE MANAGEMENT

### DESCRIPTION OF THE WORK

Hazardous waste management specialists may be employed by government or by large industries. They concentrate on understanding all aspects of hazardous waste management, including: the federal RCRA law and equivalent state laws and regulatory programs; hazardous materials chemistry and physics; facility hazardous waste compliance audit procedures; testing and classification procedures for hazardous waste; how to inspect hazardous waste generating facilities for proper drum and tank labeling and transportation manifesting; personal protection procedures and equipment; mandated treatment standards before land disposal; plant personnel training; and spill contingency plans.

Government employed hazardous waste management specialists carry out the mandates of the law, assist both large and small quantity generators in achieving compliance, and propose and draft regulations to improve the program's effectiveness in protecting human health and the environment. They inspect points of waste generation, drums and tanks used to collect and store hazardous waste, manifest records, hazardous waste storage areas, training records, hazardous waste handling, and spill response procedures. After an inspection, these specialists may issue notices of violation for serious noncompliance findings, or offer advice on correcting minor offenses.

Hazardous waste management specialists working for industry usually work at the corporate level, although one may be assigned to a plant that generates a particularly large volume of hazardous waste. Their primary activity is to advise industrial facilities on what actions to take to achieve compliance, and then oversee that these recommendations are funded and implemented. They also negotiate with regulatory agencies on compliance actions and schedules.

## For More Information

Air and Waste Management Association
Box 2861
Pittsburgh, PA 15230
(412) 232-3444

## *Program Specialist: Surface Water Quality*

## Description of the Work

Surface water quality specialists work to protect the quality of water in streams, rivers, lakes, estuaries, and oceans. Much of their work is involved with carrying out, or complying with, the federal Clean Water Act or related state legislation. Developing a sound working knowledge of regulatory standards and procedures, maintaining a familiarity with wastewater treatment technologies, and understanding the dynamics of stream or marine ecology are important responsibilities for these environmental professionals.

Surface water quality specialists are concerned with the treatment of municipal and industrial wastewater before it is discharged into a stream. They get involved in inspecting wastewater treatment plants, monitoring the quality of wastewater being discharged, reviewing sampling data to ensure that the requirements of a discharge permit are being met, and identifying and correcting problems.

Surface water program specialists employed by federal or state government agencies assist in preparing discharge permits, help set limitations on the amount of particular pollutants that may be discharged, issue notices of violation, and administer funds to local agencies for treatment system construction and upgrades. Some also conduct special studies, such as how a specific toxic chemical effects freshwater fish and plant life. Field work for surface water quality specialists often involves the planning and implementation of stream surveys. Water, sediment, fish, and/or plant samples are taken along designated stream segments and then analyzed to determine any adverse impacts caused by pollution. From such survey data, these program specialists can develop stream models to predict the water quality changes attributable to various types and levels of pollutants.

Industrial employers hire surface water quality specialists to oversee the treatment and discharge of industrial wastewater from their plants. Their duties might include: conducting internal audits of treatment systems and procedures; setting up monitoring stations; preparing monitoring reports for submittal to government agencies; planning expansions and upgrades to treatment systems; investigating spills and equipment breakdowns; coordinating special studies being performed by staff or consultants; and negotiating with government regulators on special conditions for a discharge permit.

### For More Information

Water Pollution Control Federation
601 Wythe Street
Alexandria, VA 22314-1994
(703) 684-2400

## *Program Specialist: Ground water Quality*

### Description of the Work

Ground water quality specialists work to protect the ground water from contamination, and to clean up ground water that has already been contaminated. They are especially concerned with ground water that may be used for public or private drinking water supplies. These specialists may be trained as engineers, hydrogeologists, or environmental scientists.

Both field work and administrative work are involved in the protection of ground water resources. The field work aspect of a ground water specialist's duties involves: investigating citizen complaints of possible contamination; sampling drums, tanks, soil, and ground water; interpreting the analytical results to determine levels of contamination; supervising the installation of ground water monitoring wells and the taking of soil borings; the overseeing of contractors who are excavating and removing contaminated material from a site; and responding to spills and accidents which may jeopardize ground water quality.

On the administrative side, ground water protection activities are largely related to the procedural requirements of the federal

Superfund or hazardous waste programs, or to state equivalent programs. Specific tasks include: analyzing site information to rank contaminated sites based on environmental risk; identifying parties responsible for contamination; arranging for alternative water supplies for citizens whose residential wells have been contaminated; evaluating the work of cleanup contractors; authorizing the closing of dumps and other potential sources of ground water contamination; delineating and mapping critical aquifer recharge zones to protect them from development; and coordinating communication and public relations activities for large project sites.

Government ground water quality specialists also become involved in enforcement activities, permitting, negotiating cleanup agreements, issuing violation notices, and establishing cleanup levels for contaminated sites. Ground water quality specialists employed by industry are concerned with accurately documenting and investigating site conditions; making required notifications to government agencies; evaluating the most effective and cost efficient methods for cleaning up contaminated soil and ground water; negotiating cleanup agreements; and evaluating and supervising the work of cleanup contractors.

### For More Information

National Water Well Association
6375 Riverside Drive
Dublin, OH
(614) 761-1711

Association of Ground Water Scientists and Engineers
6375 Riverside Drive
Dublin, OH 43017
(614) 761-1711

## *Program Specialist: Solid Waste Management*

### Description of the Work

Environmental specialists working in the area of solid waste management focus on the regulation of solid waste disposal facilities.

With the declining landfill capacities around the country, and spiraling costs associated with constructing new landfills, these program specialists look for alternative means for disposal. One fast growing field within solid waste management is resource recovery, where specialists concentrate on ways to reduce and reuse both municipal and industrial solid waste.

In the government sector, usually at the state level, the traditional role of solid waste management specialists was to license and regulate landfills. This is still the case today, but much more emphasis is now placed on the environmental aspects of land-fill design and operation. In addition, many states require local governments to prepare solid waste management plans which are subject to the review, evaluation, and approval of these program specialists. As part of these requirements, local agencies are requested to develop alternative means, other than landfills, to manage and dispose of solid wastes. State officials often provide advice and technical assistance on these local planning projects.

Beyond the regulatory aspects of solid waste management, resource recovery specialists promote reclamation, recycling, and reuse of solid waste materials. They become heavily involved in reviewing and evaluating the efficiencies and costs of new technologies for recycling paper, plastics, oil, metals (such as aluminum and copper), and other materials. They may also work with or for chemical manufacturers on developing programs and techniques to reclaim various chemical products, such as widely used industrial solvents. Identifying potential markets for recycled and reclaimed material is another important function of the resource recovery specialist. Often it is found that one industry's wastes can serve as another industry's raw material. Many communities are embarking on waste-to-energy projects, where municipal solid waste is burned to create steam, which in turn is used to generate electricity. Resource recovery specialists may provide advice and technical assistance to local agencies on these complex projects.

Many industrial processes generate extremely large amounts of sludge, ash, and other waste materials which traditionally have been landfilled on the plant site, incinerated, or shipped off to a local municipal landfill. Therefore, in the industrial sector, solid waste management specialists may be asked to prepare plans to ensure adequate disposal capacity for industrial solid waste, as well as to identify changes in industrial processes which would reduce the amount of solid waste produced. This is a particularly challenging task in light of new environmental regulations currently being proposed which will substantially increase the cost of industrial landfilling.

## For More Information

Air and Waste Management Association
Box 2861
Pittsburgh, PA 15230
(412) 232-3444

## *Program Administrator*

## Description of the Work

Program administrators are responsible for the day-to-day operations of government and industry environmental protection programs, such as for hazardous waste, solid waste, surface water quality, or ground water quality. This is the management level for the program specialists discussed earlier in this section. The administrative duties of these environmental professionals include: hiring, firing, and supervising staff; training staff; dealing with personnel problems; developing proposed annual budgets; proposing policy; coordinating activities with related government or corporate programs; interpreting laws and regulations; conducting public relations and education activities; scheduling and assigning work; solving problems; monitoring progress and effectiveness of programs; and identifying and implementing corrective actions.

## Educational Preparation

Environmental program specialists generally hold at least a bachelor of science degree in some environmental discipline, such as engineering, chemistry, ecology, biology, geology, environmental studies, planning, or even law. However, many employers are looking for candidates with a master's or graduate professional degree. Continuing education is critical for these program specialists in order to stay abreast of constantly changing regulations, practices, and technologies.

## Potential Employers

**Federal Government** — The majority of environmental program specialists employed by the federal government work for the Environmental Protection Agency; others may be employed in more limited numbers by the Department of Energy and the Department of Defense, including the U.S. Army Corps of Engineers.

**State Government** — State environmental protection agencies employ program specialists in virtually all of their environmental management programs.

**Local Government** — Large city and metropolitan governments may also employ environmental program specialists in their health or public works departments.

**Industry** — Large manufacturing industries with corporate environmental staffs often employ program specialists to keep the company abreast of changes in specific environmental programs and regulations. These environmental specialists usually are involved in training company managers, planning to achieve compliance with upcoming environmental regulations, and advising corporate management on ways to reduce or minimize compliance obligations (such as through the development and implementation of waste minimization programs).

# CHAPTER 4

# Careers in Environmental Health and Safety

# Careers in Environmental Health and Safety

Careers in environmental health and safety deal with the actual and potential affects of environmental conditions or circumstances on human health. Professionals in this field are particularly focused on the human health effects of environmental contaminants, and on methods to protect people from the harmful consequences of exposure to such substances.

## RISK ASSESSMENT SPECIALIST

### Description of the Work

The term "risk assessment specialist" is a functional job title that is new in the environmental job market. Environmental risk assessment is an area of study and a process for evaluating the potential of a specific environmental condition to adversely affect human health and the environment. Such assessments are required under several environmental laws to assist government regulators and decision-makers in determining the maximum amounts of chemical pollutants allowed to be discharged to the environment under a permit, and the concentrations of a chemical constituent that may remain in the soil or ground water when a contaminated site is declared "clean". Risk assessment specialists are the people who coordinate the evaluation process and pull together and prepare the risk assessment document. They coordinate teams of environmental specialists from the fields of hydrogeology, toxicology, air science, hazardous waste management, ecology, and related areas.

In conducting a human health evaluation for a site cleanup project or proposed new pollution source, risk assessment specialists perform several important tasks. Some of these tasks include: reviewing laws and regulations; identifying what information is needed; evaluating the site in question; investigating how contaminants move through the air and water; determining the ambient (i.e., background) environmental conditions; understanding how humans, plants, and animals are exposed to contaminants; comparing sample data in a contaminated area with background concentrations; determining what happens to the contaminants left in the environment; calculating human intakes of contaminants; determining how toxic or poisonous the contaminants are; estimating the strength of the chemical when humans are exposed; and calculating the intake of the chemical by humans. The point of this effort is to help government agencies decide if a proposed project or action should happen, or if the risks to human health and the environment are too great.

## Educational Preparation

Risk assessment specialists hold a minimum of a bachelor of science degree, usually in chemistry, biology, ecology, physics, or related natural or health sciences. The trend is for these specialists to hold advanced degrees in fields such as toxicology, environmental science/management, hydrogeology, environmental health, or related disciplines.

## Potential Employers

**Federal government** — Because risk assessment is more commonly being used to support decision-making in all federal environmental management programs, the Environmental Protection Agency is the primary employer of risk assessment specialists.

**State government** — Risk assessment specialists are also employed by state environmental protection agencies and, occasionally, by state departments of public health and agriculture (in relation to pesticide usage).

**Industry** — Large chemical companies are beginning to employ risk assessment specialists as they get more involved in site cleanup and other environmental management activities.

**Consultants** — Risk assessment specialists are increasingly being employed by environmental consulting firms to assist with determining cleanup standards for site remediation projects, and air emission levels for toxic chemicals.

**Academia** — The art of environmental risk assessment is continually being developed and refined through government and industry supported research activities at major universities. Risk assessment specialists are employed by universities to assist in these research activities.

### For More Information

Society for Risk Analysis
8000 Westpark Drive
Suite 400
McLean, VA 22102-3101

Air and Waste Management Association
Box 2861
Pittsburgh, PA 15230
(412) 232-3444

National Association of Environmental Professionals
P.O. Box 15210
Alexandria, VA 22309-0210
(703) 660-2364

Association of Ground Water Scientists and Engineers
6375 Riverside Drive
Dublin, OH 43017
(614) 761-1711

## REFERENCE

*Risk Assessment Guidance for Superfund, Vol. I, Human Health Evaluation (Part A), Interim Final*, U.S. Environmental Protection Agency, Office of Emergency and Remedial Response, Washington, D.C. 20460, EPA/540/1-89/002, December 1989.

# ENVIRONMENTAL TOXICOLOGIST

## DESCRIPTION OF THE WORK

Toxicologists seek to determine the harmful effects of chemicals on humans and the biological environment. Combining the principles of chemistry, biology, and related disciplines, toxicologists study the relationships between chemicals and the adverse health consequences which they can cause. Also, they estimate the likelihood that adverse health effects would be caused by a specific set of circumstances as part of environmental risk assessments.

Many environmental laws govern the release and disposal of toxic chemicals into the environment, and the cleanup of sites contaminated with toxic wastes. In carrying out the mandates of these statutes, environmental toxicologists play a critical role in answering difficult questions, such as:

- What amount of toxic material should industrial facilities be allowed to discharge into the air and water? At what rates? At what level of human exposure to specific environmental toxicants do adverse health effects begin to occur?
- At what point in the cleanup of a contaminated site is the work allowed to stop? In other words, "how clean is clean?"

Many toxicologists work in research positions in academia, industry, and government to develop new knowledge. Among the specialty areas of research in toxicology are chemical carcinogenesis, reproductive and developmental toxicology, neurotoxicology, immunotoxicology, and inhalation toxicology. Research may be "pure", which adds to the scientific base of toxicological knowledge, or "applied", where some direct social benefit is expected to be achieved. An example of the former would be studies of how a particular chemical alters the reproductive capacity of a certain type of cell. An example of the latter would be studies determining if human exposures to chemicals emitted from a particular industrial facility are high enough to be of concern to residents in the surrounding community.

In academic settings, toxicologists conducting research also teach courses at both the undergraduate and graduate levels. Increasing numbers of colleges and universities are adding toxicology courses and curricula. Because the field of toxicology is growing rapidly, opportunities for teaching in this field are expected to increase as well.

Chemical, pharmaceutical, and other industries employ many toxicologists to evaluate the safety of their products. Federal laws often require the manufacturers of drugs, food additives, cosmetics, pesticides, and other chemicals to rigidly test their products before they can be released to the commercial market. It is the responsibility of toxicologists to ensure that such testing is carried out in a scientifically sound and defensible manner. These studies are reviewed by regulatory agencies, such as the Food and Drug Administration and the Environmental Protection Agency, to make sure new products pose no unreasonable risk to human health or the environment.

## Educational Preparation

Traditionally, people performing environmental toxicology studies have held degrees in biology, chemistry, engineering, and pharmacology, without much specialized training in toxicology. However, with many colleges and universities developing programs in toxicology or adding toxicology courses to biology, chemistry, and engineering programs, students that receive such specialized training will have a distinct advantage.

Although some industries and research institutions employ toxicologists with only a bachelor's degree (usually as research assistants or laboratory workers), those with master's or doctoral degrees will find the most opportunities in government, industry, or consulting firms. The doctoral degree is generally required for lead research positions.

Graduate level programs in toxicology usually require specific prerequisites for admission. Advanced course work in chemistry, biology, microbiology, biochemistry, physiology, physics, and mathematics (including calculus and statistics) all help to provide students with a solid background for such graduate level study.

Because toxicologists are often called upon communicate with the public and government agencies, effective communication skills, both in writing and public speaking, are important for the prospective toxicologist to develop.

## Potential Employers

**Federal government** — Most environmental toxicologists working for the federal government are employed by the Environmental Protection Agency or the various agencies of the Department of Health and Human Services (i.e., the Agency for Toxic Substances and Disease Registry, the Centers for Disease Control, and the National Institutes of Health).

**State government** — States employ environmental toxicologists in their departments of environmental protection and public health to evaluate the toxicity of pollutants and contaminants.

**Consulting** — Environmental consulting firms employ environmental toxicologists to assist with risk assessments at cleanup sites, permitting air and water discharges, and the development of chemical safety and emergency response plans for government and industry clients.

**Industry** — Large chemical companies often employ toxicologists to evaluate and reduce the toxicity of their products to protect humans and the environment.

**Research institutions** — Environmental toxicologists are employed by universities and research institutes conducting environmental studies on chemicals.

## For More Information

Society of Toxicology
1101 Fourteenth Street N.W.
Washington, D.C. 20005
(202) 293-5935

## REFERENCES

*1990-1991 Careers in Toxicology* (brochure), The Society of Toxicology, 1101 Fourteenth Street N.W., Washington, D.C. 20005.

*Resource Guide to Careers in Toxicology*, The Society of Toxicology, November 1989, p. 5-9.

*Risk Assessment Guidance for Superfund Vol. I, Human Health Evaluation Manual (Part A), Interim Final*, U.S. Environmental Protection Agency, Office of Emergency and Remedial Response, Washington, D.C. 20460, EPA/540/1-89/002, December 1989.

## RISK COMMUNICATION SPECIALIST

### Description of the Work

Risk communication has recently evolved as a specialty area associated with community and public relations activities. Specialists in risk communication use their technical knowledge of environmental issues, combined with well-developed writing and public speaking skills, to interpret and present complex scientific issues for general citizen understanding.

Environmental risk is a difficult concept to convey to others because some level of comprehension of the technical work of environmental engineers, hydrogeologists, toxicologists, risk assessment specialists, regulatory compliance experts, and environmental scientists is required. Risk communication specialists are technical generalists who can speak the language and jargon of the technical specialists, and are able to decipher the critical facts for the layman.

The ultimate objective of risk communication is to facilitate public understanding of environmental risks to the point where citizens can make informed personal decisions about how to respond to such risks. In carrying out this mission, risk communication specialists: work to understand public perceptions of risks associated with a particular project or action; develop mechanisms for citizens to provide constructive feedback to industrial or governmental decision-makers; find effective ways to simplify and present complex technical concepts to public groups while maintaining an appropriate level of scientific accuracy; train and coach agency or industry spokespersons in risk communication concepts; prepare risk communication plans for controversial, or potentially controversial, environmental projects or actions (such as the siting of a hazardous waste incinerator); prepare and implement strategies to release critical environmental risk information to the public and the news media as soon as it becomes available (such as during a remedial investigation at a Superfund site); and coordinate educational seminars to provide citizens an opportunity to become familiar with difficult-to-understand scientific data and terminology.

Listening and responding in a meaningful way to citizen concerns about issues of environmental risk is essential to building trust between citizens and project sponsors, regardless of whether they represent government or industry. This is a key role, and perhaps the most valuable one, for the risk communication specialist. If the public loses trust in a project sponsor, the resulting outrage can lead to bad publicity, lawsuits, and can even stop a project unnecessarily. Additionally, when project sponsors lose their credibility with the public, the perception of environmental risk posed by a project or action is driven by emotion, rather than facts, which often leads to an exaggeration of the actual risk.

## Educational Preparation

Risk communication specialists should have a sound technical background in an environmental discipline, such as environmental engineering, ecology, hydrogeology, or environmental health. To have the most career flexibility, aspiring risk communication specialists should have at least a general understanding of several environmental disciplines (i.e., biology, geology, hydrology, toxicology, meteorology, engineering, etc.). A bachelor of science degree in a technical environmental field would be a minimum requirement for this profession. Those with master's and doctoral degrees will probably have more opportunities.

In addition, sound communication skills are required. College electives in English, writing, interpersonal communications, public speaking, and journalism would be very helpful to candidates for risk communication positions.

To further round out this educational background, additional course work in the social sciences (i.e., sociology, psychology, political science, and economics) would provide students with insight into how and why citizens may respond to environmental controversies.

## Potential Employers

**Federal government** — Risk communication specialists are employed by the federal government at the Environmental Protection

Agency, Department of Energy, and other agencies that get involved in environmental controversies. Most are employed in the Superfund program during the investigation and cleanup of contaminated sites.

**State government** — Most states assist in carrying out the federal mandates of Superfund, as well as administering their own site cleanup programs. Risk communication specialists are employed by states to implement and coordinate communication at these sites, as well as to assist in other controversial actions, such as issuing air toxics permits to industrial facilities.

**Consulting firms** — Risk communications specialists are occasionally employed by engineering and environmental consulting firms to assist industry and local governments with communications during controversial environmental projects.

**Public relations firms** — As public relations firms become increasingly involved in environmental controversies, they will need to employ risk communication specialists to assist with the interpretation of technical information.

**Industry** — Large manufacturing industries occasionally employ risk communication specialists to assist in communicating technical information during environmental controversies.

### For More Information

National Association of Professional Environmental Communicators
P.O. Box 06 8352
Chicago, IL 60606-8352
(312) 321-3336

## SANITARIAN/ENVIRONMENTAL HEALTH SCIENTIST

### Description of the Work

Sanitarians, or environmental health scientists as they are sometimes called, work to protect the health, safety, and comfort of the public. One of their primary functions is to inspect buildings, facilities, equipment, and property to assure compliance with public health codes and related environmental regulations. Among the inspection sites are hospitals, restaurants, homes and apartments, industrial plants, municipal landfills, water supply facilities, wastewater treatment plants, schools, supermarkets, food processing plants, and camps and recreation facilities. During these inspections, sanitarians check water supplies, waste disposal facilities, sources of air and noise pollution, signs of unsanitary conditions (rats, mice, general filth, spilled material, etc.), food preparation practices, cockroaches, restrooms, and other types of health hazards.

These environmental health scientists perform percolation tests on soils where construction is proposed that will use septic systems and drain fields to treat and dispose of sewage. They also respond to citizen complaints about water coming from private wells or community water supply systems, by taking water samples and sending them to a laboratory for analysis. Among some of the other duties of sanitarians are the inspection of public swimming pools to assure proper chlorination and management of bacteria; working with dairy farmers and food producers to make sure proper handling procedures are followed; providing instruction to apartment building and restaurant owners on safe sanitation practices; checking on hospitals using radioactive isotopes to see if the low-level radioactive waste is disposed of according to applicable regulations; and developing and implementing programs to deal with incidental public health problems such as an outbreak of a disease carried by a certain breed of mosquito.

Environmental health scientists, in conjunction with emergency response agencies, assist in developing plans to protect public health during both natural (floods, tornados, earthquakes, etc.) and man-made emergencies (industrial explosions, spills of hazardous materials, chemical tank car derailments, etc.). In the latter case (such as a chemical emergency), they help to decide when neighborhoods should be evacuated, where evacuees should go for shelter, and how to provide safe drinking water and waste disposal during the emergency.

During the course of their work, sanitarians may be asked to gather evidence to support legal actions, issue warnings or citations to health code violators, conduct special studies on various environmental health issues to determine the extent of a problem and to identify practical and cost-effective solutions, testify in court proceedings, work with the local news media, and attend training seminars to keep up-to-date on regulations and management techniques.

Most sanitarians work for state and local health agencies. Those working for local government in large cities may have the opportunity to specialize in one field, such as water quality, food preparation, or solid waste management, while those in rural areas may have to coordinate an entire environmental health program for a governmental jurisdiction.

## Educational Preparation

A bachelor of science degree in biology, chemistry, engineering, or environmental health is the minimum level of education for environmental health scientists. Advanced courses in microbiology, biostatistics, epidemiology, health education, and public administration would be helpful for the prospective sanitarian. Beyond a core curriculum in science, students should develop their communications skills by taking courses in writing and public speaking. Also, social science courses, particularly sociology and psychology, are helpful in learning to deal with people under stressful circumstances. Those desiring to advance into management positions should earn a master's degree in public or environmental health. A doctoral degree prepares a student for research and teaching positions, as well as for top management positions in governmental agencies.

Sanitarians are required to be registered in many states. The registration requirements usually involve one to two years of experience and the passing of a registration examination.

## Potential Employers

**Federal government** — Environmental health scientists are employed in the federal government by the Department of Health and Human Services (Public Health Service), the Environmental Protection Agency, and occasionally by other agencies such as the Food and Drug Administration and the Department of Defense.

**State government** — States employ sanitarians in their departments of public health and environmental protection.

**Local government** — Sanitarians are employed by city, county, metropolitan, and regional health departments.

## For More Information

National Environmental Health Association
720 South Colorado Blvd.
South Tower, Suite 970
Denver, CO 80222
(303) 756-9090

Society for Occupational and Environmental Health
6728 Old McLean Village Drive
McLean, VA 22101
(703) 556-9222

## REFERENCE

*Sanitarians (Environmental Health)*, Brief 431, Chronicle Guidance Publications, Inc., Moravia, NY 13118, March 1989.

# EPIDEMIOLOGIST

## Description of the Work

Epidemiology is the study of diseases and epidemics as they affect human populations. Epidemiologists working in the environmental field study human populations to identify and validate correlations between a particular set of environmental factors and specific diseases or health conditions. These studies attempt to link human exposure to a particular toxic chemical with an identified chronic health effect, or a combination of effects.

Examples of major successful epidemiological studies include those that: linked smoking with lung cancer and other diseases; linked asbestos with asbestosis and other chronic ailments; linked vinyl chloride with a rare type of liver cancer; and linked benzene with leukemia. Because of the long list of toxic chemicals associated with Superfund and state-listed contaminated sites (which number in the thousands), and with new regulatory toxic air and water discharge limits, increasing numbers of epidemiological studies will likely be conducted to determine the long-term chronic health effects of exposure.

In a full-scale epidemiological study, epidemiologists set out to test a scientific hypothesis about some cause and effect relationships between human exposures to hazardous substances and certain chronic health effects. As part of these analytical investigations, some of the variables that are evaluated are the distribution of the affected subjects, their age, sex, occupation, and economic status. Such studies often involve two groups of subjects, one group of people who have been exposed to the substances in question, and another group who have not been exposed (i.e., the control group). For the duration of a study, which may take several years or even decades, epidemiologists coordinate the collection and analysis of exposure and health information provided by the study populations. In their analysis, they attempt to statistically relate exposures and disease.

## Educational Preparation

Epidemiologists are essentially public health researchers, and as such usually hold graduate level degrees in disciplines such as environmental health, medicine, sociology, toxicology, statistics, and survey research.

## Potential Employers

**Federal government** — Epidemiologists employed by the federal government work for the Agency for Toxic Substances and Disease Registry, the Centers for Disease Control, the National Institutes of Health, the Environmental Protection Agency, and the Food and Drug Administration.

**State government** — States employ epidemiologists in their departments of public health.

**Industry** — Epidemiologists may be employed by the larger chemical and pharmaceutical companies.

**Research institutions** — Research institutions affiliated with medical centers, universities, and industry employ epidemiologists to study human diseases and conditions which may be attributable to environmental conditions.

### For More Information

Society for Epidemiologic Research
c/o American Journal of Epidemiology
2007 East Monument Street
Baltimore, MD 21205
(301) 955-3441

## REFERENCES

Kamrin, M. A., *Toxicology — A Primer on Toxicology*, Lewis Publishers, Inc., 121 South Main Street, Chelsea, MI 48118, 1988, p. 53.

*Environmental Regulatory Glossary, Fifth Edition*, Frick, G. W. and Sullivan, T. F. P., eds., Government Institutes, Inc., 966 Hungerford Drive, #24, Rockville, MD 20850, March 1990, p. 139.

*Risk Assessment Guidance for Superfund, Vol. I, Human Health Evaluation Manual (Part A), Interim Final*, U.S. Environmental Protection Agency, Office of Emergency and Remedial Response, Washington, D.C. 20460, EPA/540/1-89/002, December 1989, p. 2-10.

## EMERGENCY RESPONSE SPECIALIST

### Description of the Work

Emergency response specialists are responsible for planning, coordinating, and executing emergency response procedures for accidents and spills involving hazardous substances. They may be employed by government agencies, industries, specialty contractor firms, or consulting companies.

Working with the specific knowledge of the hazardous substances used or produced at a site or in a community, emergency response specialists prepare detailed instructions for responding to a chemical accident, from initial reporting of a spill, identifying the material, and implementing safety precautions to protect employees and/or citizens, to initiating containment actions and calling in trained fire-fighters and cleanup teams.

Several federal, state, and local laws and regulations mandate that communities and industries develop hazardous chemical emergency plans. Examples include: the Spill Prevention, Contingency and Countermeasures Plan under the Clean Water Act, the Contingency Plan requirement for large quantity hazardous waste generators under the Resource Conservation and Recovery Act, and the requirement that Local Emergency Planning Committees prepare facility-specific emergency response plans under Title III of the Superfund Amendments and Reauthorization Act.

Emergency response specialists are assigned the task of developing these complex plans, and must collect a substantial amount of information: facility layout drawings, an inventory of all emergency response equipment, plant operating procedures, identified on-call cleanup contractors, training records for employees participating in emergency response activities, list of local emergency agencies which can be called upon in extreme emergencies, inventories of hazardous substances used or produced at the site, arrangements with local hospitals and medical facilities, current alarm systems, evacuation routes, and documentation of emergency response exercises which have been conducted. These planning efforts are often coordinated with state and local police departments, local fire departments, and other potentially participating parties.

Ensuring that plant employees are fully trained in emergency response procedures is a key responsibility of the emergency response specialist. It is critical that employees understand how to respond to chemical emergencies, from knowing what communications must take place to what types of personal protective equipment must be worn or used. In addition, citizens in the surrounding community need to be informed about how to respond to a hazardous chemical accident at a nearby plant. Emergency response specialists work closely with local agencies to prepare communities for such an accident.

## Educational Preparation

Emergency response specialists come from both law enforcement and fire-fighting backgrounds, as well as from environmental protection backgrounds. Although currently some emergency response specialists do not hold bachelor's degrees, due to the complexity and broad training required by these positions, the clear trend is that candidates for management positions hold a minimum of a bachelor's degree. A master's degree promises candidates more likely opportunities for management positions in both government and industry.

## Potential Employers

**Federal government** — Emergency response specialists are employed by the Environmental Protection Agency, the Federal Emergency Management Agency, and the U.S. Coast Guard.

**State government** — Emergency response specialists are employed by state governments in their departments of environmental protection and state police, and in their fire marshal offices.

**Local government** — Emergency response specialists are employed in local police departments, fire departments, and health departments. Some communities employ emergency response specialists to coordinate specially trained hazardous materials response teams.

**Industry** — Large manufacturing industries, such as chemical and petroleum companies, employ emergency response specialists to coordinate spill response teams at production facilities.

## For More Information

National Emergency Management Association
c/o SEMA
P.O. Box 116
Jefferson City, MO 65102
(314) 751-9571

Spill Control Association of America
400 Renaissance Center
Suite 1900
Detroit, MI 48243
(313) 567-0500

Chemical Waste Transportation Institute
1730 Rhode Island Avenue N.W.
Suite 1000
Washington, D.C. 20036
(202) 659-4613

Chemical Manufacturers Association
Community Awareness and Emergency Response Program
2501 M Street N.W.
Washington, D.C. 20037
(202) 887-1100

Other good sources are the U.S. Environmental Protection Agency and the U.S. Coast Guard:

U.S. Environmental Protection Agency
Office of Emergency and Remedial Response
Emergency Response Division
401 M Street, S.W.
Washington, D.C. 20460

U.S. Coast Guard
Office of Marine Safety, Security, and Environmental Protection
Marine Environmental Response Division
2100 Second Street, S.W.
Washington, D.C. 20593

## INDUSTRIAL HYGIENIST

### Description of the Work

Industrial hygienists work to protect the health of employees and to improve the conditions of the working environment. More specifically, they manage and control environmental factors and stresses in the workplace which may adversely affect workers or citizens in the community. If not managed and controlled effectively, poor workplace conditions may lead to worker sickness, discomfort, and/or inefficiency. The science and art of industrial hygiene involves the recognition of environmental hazards and sources of stress, the evaluation and measurement of those factors, and the prescribing of methods and procedures to control or eliminate them.

In working to improve the work environment, industrial hygienists perform a variety of related tasks, from studying workplace hazards and worker exposure levels to proposing protective measures and establishing standards of safety. Examples of the types of environmental hazards found in the workplace include: indoor air particulates (such as aerosols, dusts, and asbestos fibers); biological agents (plants, animals, fungi/yeast/algae microorganisms, and parasites); industrial process noise; ionizing (alpha, beta, or gamma) and nonionizing (such as ultraviolet and infrared) radiation; and gas and vapor hazards from fugitive emissions of volatile organic compounds from industrial equipment or other sources.

To protect workers from environmental hazards, industrial hygienists recommend various types of protective equipment and working procedures. For example, an industrial hygienist may recommend or require that a worker handling a hazardous chemical wear a chemically resistant suit, rubber gloves, safety glasses, and a respirator. In addition, it may be required that a specially designed forklift for handling drums of hazardous waste be used, and that detailed drum handling procedures be followed. Another example involves employees that work in confined spaces with a vapor hazard, such as inside a tank or pipeline, where industrial hygienists would specify detailed safety procedures and equipment.

Industrial hygienists often work in teams with physicians, toxicologists, safety specialists, and environmental health scientists to determine the adverse affects of the workplace on workers' health (such as in an epidemiological study of workers at a nuclear power facility), to improve environmental conditions in the workplace, and to protect the community. Industrial hygienists often participate in emergency response, planning activities to help determine levels of exposure to workers and citizens of a particularly toxic chemical used at a plant.

Aside from an employer's desire to protect workers and the community from environmental hazards associated with an industrial process, a federal law (the Occupational Safety and Health Act) mandates that certain occupational health and safety standards be followed. One section of that law establishes a hazard communication standard, commonly referred to as the "worker right-to-know" provisions. Under this regulation, employers are required to post or make available to workers information on any hazardous chemicals present in the workplace. Industrial hygienists are often assigned the task of complying with the requirements of this program.

### Educational Preparation

A bachelor of science degree in engineering, chemistry, physics, or the physical or biological sciences is the minimum educational requirement for aspiring industrial hygienists. Advanced degrees in occupational and environmental health, medicine, toxicology, or other sciences would provide additional preparation for work in industrial hygiene.

### Potential Employers

**Federal government** — Industrial hygienists are employed by the federal government in the Department of Labor, particularly the Occupational Safety and Health Administration.

**State government** — Industrial hygienists are employed as inspectors for state labor and public health agencies.

**Industry** — Industries employ the most industrial hygienists to administer health and safety programs at industrial facilities.

**Consulting firms** — Some industrial hygienists are employed as consultants to government and industry.

### For More Information

American Industrial Hygiene Association
P.O. Box 8390
345 White Pond Drive
Akron, OH 44320
(216) 873-2442

## REFERENCE

American Industrial Hygiene Association, Membership Directory 1990-1991.

# HEALTH PHYSICIST

## Description of the Work

Health physicists are concerned with protection of workers, communities, and the environment from the harmful effects of exposure to radiation. These professionals may specialize in applications involving nuclear power generation, medicine, research, industrial processes, regulatory enforcement, environmental radiation (especially radon), nuclear waste management, and remedial actions (i.e., the cleanup of contaminated areas). The identification, understanding, prevention, and control of radiation hazards is the primary focus of work for health physicists.

Health physicists are involved in several important technical activities related to radiation protection and control. Examples of the work include: reviewing radiological health monitoring data of employees; developing emergency response procedures for radiation accidents; establishing site-specific safety standards for radiation control and personal protection; and training of radiation workers in personal protection methods. Health physicists working for government agencies assist in developing and enforcing radiation control regulations, as well as inspecting of facilities using radioactive materials. They may also be called to assist architects and engineers in designing structures to contain radioactive materials, or to assist in the design of radiation detection equipment. One growing area of involvement for health physicists is in assessing the environmental impact of radiation releases.

The demand for health physicists is increasing, and even accelerating, as radioactive materials are being more commonly used in a variety of industrial, governmental, and medical applications. Particularly in the environmental field, health physicists are needed to assist in the cleanup of U.S. Department of Energy sites across the country where soils, ground water, and waste materials have been radioactively contaminated. Prospects for careers in health physics will remain excellent well into the next century.

## Educational Preparation

Health physics, as the science of radiation control, is interdisciplinary in nature. It relates closely to several other scientific disciplines, such as biology, ecology, chemistry, physics, nuclear engineering, medicine, physiology, and toxicology. These professionals need to develop a broad scientific background in order to understand the complexities of such things as how radioactive isotopes interact with matter, the nature of radioactive environments, and the impacts of radiation on life processes.

The minimum level of education for a professional health physicist is a bachelor's degree in the sciences, with course work in biology, chemistry, physics, and mathematics. Some specialized courses in nuclear engineering, radiological health, and radiation biology will prove very useful to the prospective health physicist. For those specifically interested in environmental issues, additional elective course work in environmental engineering and hydrogeology will be valuable. Many universities are now offering specific bachelor's level programs in health physics.

Many health physicists continue with their education to obtain a master's or doctorate degree. As with most scientific professions, graduate degrees are required for most teaching and research positions.

The field of health physics also offers employment opportunities for those trained as technicians. Several colleges offer two-year technical degrees in health physics, which include training in radiation survey and analytical techniques, instrument calibration and operation, and monitoring.

## Potential Employers

**Federal government** — Health physicists are employed by the Nuclear Regulatory Commission, the Department of Energy, the Department of Defense, the Environmental Protection Agency, the Department of Health and Human Services, and the Department of Labor.

**State government** — States employ health physicists in their health, labor, and/or environmental protection departments.

**Local government** — Large city or metropolitan health departments employ health physicists as inspectors and advisors to municipal power utilities using nuclear reactors.

**Industry** — Several types of companies employ health physicists, including utility companies which own and operate nuclear power plants, nuclear fuel fabrication companies, protective equipment manufacturers, and nuclear engineering and consulting firms.

### For More Information

Health Physics Society
8000 Westpark Drive
Suite 400
McLean, VA 22102
(703) 790-1745

## REFERENCES

"Health Physics, The Profession of Radiation Protection", a brochure prepared by the Health Physics Society, rev. 7/89.

"Health Physics, A Profession of the Nuclear Age", a pamphlet prepared by the Midwest Chapter of the Health Physics Society, no date.

CHAPTER 5

# Careers in Environmental Education

# Careers in Environmental Education

Many career paths exist for those individuals interested and skilled in teaching others about environmental subjects. For environmental education professionals, the primary reward is job satisfaction through continual personal contact with students, seminar participants, citizens, readers, and viewers. This chapter will briefly survey these career opportunities.

## SCHOOL TEACHER

Elementary, junior high, and high school teachers in both public and private schools have a unique opportunity to assist children in developing a respect for natural processes and to instill an awareness of environmental values. However, school teachers seldom get the professional opportunity, in practice, to specialize in the environmental education area (although education majors in college can often specialize in environmental education in their undergraduate or graduate curricula). Most environmental education teaching is conducted as part of other subjects, such as biology, geography, chemistry, social studies, civics, etc.

### Description of the Work

School teachers introduce students to environmental ideas and concepts through educational programs involving: classroom lectures and discussions, classroom experiments, field studies in the school yard, guest speakers, audiovisual materials, tabletop demonstrations, and field trips. For environmental field trips, teachers often take students to nature centers, botanical gardens, zoos, museums of natural history, aquariums, parks, and hands-on science museums. For older students studying man-environment relationships, teachers may arrange field trips to game farms and wildlife refuges, fish hatcheries, wetland habitats, municipal sewage treatment plants and landfills, and industrial plants (including, perhaps, paper or aluminum recycling plants).

### Educational Preparation

To become an environmental education teacher, the same college level educational preparation is required as for any teacher. In addition, however, teachers interested in environmental education should take elective classes in subjects like ecology, chemistry, geology, biology, natural resources management, environmental engineering, and environmental science.

### Potential Employers

**Public and private school systems** — Environmental education teachers are usually employed as science teachers in public and private school systems.

### For More Information

North American Association for Environmental Education
P.O. Box 400
Troy, OH 45373
(513) 339-6835

American Society for Environmental Education
1200 Clay Street #2
San Francisco, CA 94108
(415) 931-7000

## INTERPRETIVE NATURALIST

### Description of the Work

Interpretive naturalists teach natural history subjects to school children and adults, usually as part of a nature center operation. Using classroom lectures, see-and-touch displays, aquariums, terrariums, nature hikes, slide shows, movies, and other techniques, interpretive naturalists attempt to get people interested in the workings of nature.

One of the interpretive naturalist's job responsibilities is to develop, or to coordinate the development of, displays and presentation programs. This may involve: collecting rocks, bird nests, tree leaves, and plant and animal specimens for displays; taking photographs and researching presentation material; supervising the construction and installation of interpretive signs along nature trails; and coordinating the work of consultants and contractors in designing and constructing museum-type displays.

Usually employed by local, state, and national park and recreation agencies, interpretive naturalists work at nature centers, often located in natural surroundings such as public parks or forests. Interpretive naturalists may also be employed by zoos, aquariums, arboretums, or other educational facilities. In the off season, naturalists may visit school classrooms and civic organizations to present special programs. They may, on occasion, be approached by reporters from newspapers, radio, or television to report on current happenings at their facilities.

## Educational Preparation

The minimum education requirement for a professional interpretive naturalist is a bachelor's degree in science. Beyond science classes, prospective interpretive naturalists should also take courses in public speaking, composition, and art or design.

## Potential Employers

**Federal government** — Interpretive naturalists are employed by federal agencies, such as the National Park Service and the U.S. Forest Service, which operate visitor centers, museums, nature centers, and related facilities.

**State government** — State conservation agencies employ interpretive naturalists to administer programs at state parks and recreation areas.

**Local government** — Interpretive naturalists are employed by city, county, and metropolitan park and recreation agencies to administer programs at local nature centers.

**Industry** — Certain large industries, such as electric power utilities, employ interpretive naturalists to operate visitor information centers on company-owned lands.

**Nonprofit organizations** — Nonprofit organizations that operate museums, zoos, and aquariums employ interpretive naturalists to administer their public educational programs.

**Commercial attractions** — Increasingly, large commercial tourist attractions (such as Walt Disney World in Orlando, Florida) are including environmental educational facilities. With this trend, interpretive naturalists will likely be employed to operate and manage such facilities.

### For More Information

National Association for Interpretation
P.O. Box 1892
Fort Collins, CO 80522
(303) 491-6434

## ENVIRONMENTAL TRAINER

A relatively new career that has evolved over the last ten years is the technical environmental trainer. Corporations, industrial plants, and even government agencies must stay up-to-date on a plethora of environmental regulations, pollution control technologies, spill control procedures, and environmental health and safety practices. In addition, federal hazardous waste and occupational safety regulations require annual training for certain industrial plant workers and for personnel working on contaminated sites. To meet the growing need for training, government and business organizations are beginning to retain the services of environmental consultants or environmental training companies.

## Description of the Work

Environmental trainers must have strong technical backgrounds in their area of teaching (e.g., air toxics regulations, wastewater treatment, chemical emergency response, etc.). This technical knowledge is transferred to trainees through the use of instructional techniques particularly suited for adults. These include: audiovisual materials, lectures, desktop exercises, group interaction, role playing, guest speakers, demonstrations, and other techniques. The biggest challenge for environmental trainers is to effectively relate to students of varying backgrounds and sophistication, while accurately presenting complex technical information in ways that are interesting (and even entertaining!).

## Educational Preparation

Environmental trainers should have a well-developed technical background, preferably a degree in environmental science or engineering. They should have a working knowledge of key environmental laws and regulations. College and, later, professional development training should include courses in communication, instructional techniques, and public speaking.

## Potential Employers

**Federal and state government** — Federal and state government agencies do not usually hire environmental trainers per se. However, they do employ training coordinators to arrange for technical training and professional development opportunities for environmental professionals.

**Consulting firms** — Environmental consulting firms offer regulatory and technology related training services to their governmental and industrial clients, along with their scientific and engineering services. Environmental trainers within consulting firms usually evolve out of the ranks of scientists and engineers (those who exhibit an interest and aptitude for teaching).

**Environmental training firms** — These firms specialize in developing and conducting environmental training programs, either tailored to the needs of particular clients, or offered to the public at large. They specifically employ environmental trainers with sound technical backgrounds and instructional experience.

**Academia** — As part of community education or other outreach programs, colleges and universities frequently sponsor technical environmental training seminars for students and practicing professionals. Although faculty members are usually involved in the presentation of material, these institutions occasionally employ environmental trainers to develop programs, prepare background materials, create audiovisual aids, promote the seminars, arrange logistical details, and assist in the presentations.

### For More Information

National Environmental Training Association
2930 East Camelback Road, Suite #185
Phoenix, AZ 85258
(602) 956-6099

## COMMUNITY ACTIVIST

### Description of the Work

Community activists organize citizens in response to some specific environmental issue which has a direct impact on the community. In order to gain members for the citizen organization, win sympathy for the cause within the community, and wield influence with local and state (and even federal, on occasion) legislative and regulatory bodies, the community activist must be an effective educator.

Community activists sponsor public information meetings to present the issues of concern, whether advocating the initiation of a community-wide recycling program or challenging a local pulp mill to reduce its discharge of dioxin into a nearby river. They most often utilize teaching techniques such as lectures, discussion, question and answer sessions, and a variety of visual presentation methods (slide shows, overhead transparencies, easel pad, chalkboard, posters, etc.). In addition, community activists often invite guest speakers who are perceived as experts in the subject at hand, as well as government officials who either hold relevant regulatory authority, or who administer grant programs for which the citizen organization may be eligible.

No particular education or background is required for the community activist, as many of these positions begin as unpaid volunteers. Once a citizen organization becomes well established and becomes involved in fund raising activities, the leader's position (and sometimes a few staff positions) can be compensated. Some universities have curriculums in environmental advocacy, which provide for both a scientific background and skills to facilitate social change. For a single issue organization, however, the community activist's interest and enthusiasm are the primary prerequisites for effectiveness.

### Educational Preparation

A college education is helpful but not required. Communication skills, particularly writing and public speaking, are vital.

### Potential Employers

**Nonprofit community organizations** — Community activists usually begin their careers by volunteering for a nonprofit community organization, or establishing one focused on a particular environmental issue.

## COOPERATIVE EXTENSION AGENT

### Description of the Work

Through the nation's system of land grant colleges and universities, each state has an agricultural cooperative extension service, organized to make available to the public, the knowledge and resources of the affiliated academic institution. These cooperative extension service organizations place agents in district offices throughout each state, usually located in every county. Cooperative extension agents are responsible for providing information and technical assistance to residents, and for conducting or coordinating public educational programs.

The initial role of the cooperative extension agent was to assist farmers by bringing new agricultural technologies and approaches from universities to rural areas. Over time, the educational topics addressed by cooperative extension agents have expanded. Today, several of their programs deal directly with environmental management issues. Examples include the proper use of pesticides and disposal of containers, no-till cultivation and other soil conservation techniques, ground water protection, wildlife habitat preservation, non-point source pollution control (i.e., agricultural runoff), water conservation, composting to produce fertilizer, and wastewater sludge management at livestock feedlots.

### Educational Preparation

Cooperative extension agents are employed by the affiliated academic institution. A minimum of a bachelor's degree in an environmental and/or agricultural field of study is required. Perhaps the best way to set up a career path with a cooperative extension service organization is to enroll in a land grant university, at least for undergraduate studies. This gives a student the opportunity to become acquainted with the professors and administrators of the extension service programs and to learn of job openings.

### Potential Employers

University affiliated Cooperative Extension Services.

### For More Information

National Association of County Agricultural Agents
215 City-County Building
Wheeling, WV 26003
(304) 234-3673

## COMMUNICATION ARTS PROFESSIONAL

### Description of the Work

This category of professional includes writers and video/radio program producers who create special educational features and pieces which focus on environmental issues. Whether a television documentary on protecting endangered species, or a special feature article (or series) on the solid waste dilemma for a general issue magazine, one of the objectives of these presentations is to provide an education to the public. Communication arts professionals, then, can be environmental educators through their use of the print and broadcast media.

Communications arts professionals may freelance, or be employed by a production company, publisher, or other media organization. The creative and technical prerequisites are those that involve putting the story together (conceptualizing, writing), determining the best creative and technical mechanisms for conveying the story (audio, visual, and/or print), and, finally, producing the piece. Those who specialize in environmental subjects may develop capabilities in and familiarity with these areas through reading, taking classes, and interviewing scientists, engineers, and other environmental professionals.

Many professional writers concentrate exclusively on writing about environmental topics or issues. Freelance writers may create feature stories on a broad environmental issue, such as the causes and impacts of global warming, or on a site or area related matter, such as the loss of wetlands along the Chesapeake Bay. Writers employed by magazines usually receive assignments focused on environmental topics of current controversy or interest, for example, the Exxon Valdez oil spill or a chlorine release at a city's wastewater treatment plant. Industrial trade associations employ writers knowledgeable about environmental matters to prepare articles on regulatory changes about to affect a particular industry, or about a recycling success story at a manufacturing plant. Environmental professional associations hire editors and writers to coordinate production of journals and other publications. On the artistic side, a growing number of writers are preparing fiction stories for television and movies with underlying environmental themes.

Environmental writers gather information for their stories through extensive topical research, literature searches of recent material, interviews with representatives from government, industry, academia, and the community, observation, photographs, and other means. Those writers with some technical scientific background in environmental subjects will have a significant advantage in pulling complex stories together. They will be able to ask the right questions of the right sources to help build an accurate and complete environmental story.

## Educational Preparation

Communication arts professionals come from a wide variety of educational backgrounds, with many holding degrees in English, art, and communications. To work effectively on environmental subjects, communications arts professionals should develop some scientific background through courses, seminars, and/or extensive reading.

## Potential Employers

**Magazine and book publishers** — Publishers employ editors and writers to assist in producing articles and books.

**Television stations** — Television stations, as part of either their news, special feature, or documentary programming, occasionally employ scientific or environmental advisors to assist in writing and producing stories.

**Production companies** — These companies produce documentaries and special feature programs, and employ advisors, writers, and producers familiar with environmental subjects.

**State conservation agencies and organizations** — Many state departments of conservation or natural resources publish magazines and books on environmental issues pertaining to their state. State conservation or environmental organizations often produce similar materials. Environmental editors and writers are regularly employed by these agencies and organizations.

**Environmental interest groups** — National environmental interest groups often employ communication arts professionals to write and produce educational programs for television, radio, and the print media, including their own publications.

**Self-employment** — Many communication arts professionals, particularly writers, are self-employed and work on a freelance basis with media companies.

## For More Information

Outdoor Writers Association of America
2017 Cato Avenue
Suite 101
State College, PA 16901
(814) 234-1011

National Association of Professional Environmental Communicators
P.O. Box 06 8352
Chicago, IL 60606-8352
(312) 321-3336

CHAPTER 6

# Careers in Allied Environmental Professions

# Careers in Allied Environmental Professions

The careers presented in this chapter can actively and directly involve environmental work, but are distinct professions in their own right. Generally, these types of careers require less formal education in the environmental sciences and related technical subjects, but may still require rigorous academic or creative training. Although not thought of as traditional environmental protection or natural resources management careers, individuals working in these allied environmental professions can make significant contributions to the field.

The purpose for discussing these allied environmental professions is to attempt to be inclusive of all career options available to students and career changers. Individuals interested in these careers should consult other references for more detailed information on educational requirements.

## ACTIVIST AND LOBBYIST

### DESCRIPTION OF THE WORK

Environmental activists attempt to raise public awareness about critical conservation, environmental, or health and safety related issues and to influence the outcome of legislative and bureaucratic decisions. Motivated by their mission to facilitate change for the greater public good, they are often involved in organizing local or regional citizen groups around a specific issue, such as a hydroelectric dam being proposed for construction on a designated wild and scenic river, or the proposed loosening of a pollution control regulation. Activists also coordinate the activities of special interest environmental organizations, and often get involved in writing articles for magazines and newspapers, orchestrating letter writing campaigns to legislators or congressmen, preparing and delivering testimony at public hearings, recruiting new members for an environmental interest group, planning and staging special fundraising events, conducting research on critical or controversial issues, drafting position papers and policies for an interest group, and discussing controversial issues with each of the involved parties.

Lobbyists, on the other hand, represent the interests of their clients before elected government officials. Their clients may be industrial groups, government agencies, conservation or environmental interest groups, or professional associations. Because lobbyists deal more directly and on a continual basis with governors, legislators, congressmen, senators, and other officials, they tend to be more involved in the political process than activists. The daily activities of a lobbyist may include: reviewing and evaluating the impact of proposed legislation on a client; providing testimony at a legislative committee meeting; discussing an elected official's position on a particular issue; building coalitions of among various elected officials and organizations in support of changes to proposed legislation; collecting data on critical resource management issues from local, state and federal agencies; analyzing public opinion on a specific issue and determining any probable impacts on legislation; and providing legislative updates to a client.

## Educational Preparation

Individuals considering a career as an environmental activist or lobbyist need to develop a broad educational background which balances scientific knowledge with sociology, psychology, political science, economics, English, and communications. In addition, students should take college level courses in environmental policy and law.

## Potential Employers

**Industrial trade associations** — The trade associations that represent manufacturing industries (chemical, pulp and paper, petroleum, mining) and resource-based industries, like commercial fishermen and loggers, employ lobbyists to represent them on environmental issues.

**Individual industrial companies** — Lobbyists are also employed by individual companies, usually large in size, to represent their specific interests on environmental issues or proposed legislation.

**Environmental interest groups** — Activists often serve as volunteers or staff to environmental interest groups, to coordinate educational campaigns for citizens, legislators, and the media.

**Federal, state, and local government agencies** — Activists are frequently solicited to serve part-time on government boards, special committees, or commissions dealing with environmental issues and controversies.

### For More Information

Contact employers directly for job descriptions and employment requirements.

## ENVIRONMENTAL LAWYER

### Description of the Work

Environmental lawyers work with legislative and judicial processes to represent client or employer interests on environmental matters. The clients may be individuals, businesses, organizations, or government agencies. Environmental lawyers also advise and counsel clients on specific courses of action. Their work often involves the researching and interpreting of laws, regulations, and judicial rulings as they relate to a particular case. An ability to write clear and precise reports and briefs about complex legal concepts and environmental issues is extremely important for these attorneys. In addition, client confidentiality, efficient and courteous client meetings, and adherence to strict rules of conduct are important elements for success as an environmental lawyer.

The types of issues that environmental lawyers become involved in largely depends on the client or employer. An attorney working for a state environmental agency (or attorney general's office) may frequently become involved in defending the government's actions in enforcing hazardous waste management environmental regulations. A nonprofit environmental interest group may have its legal counsel file a lawsuit to stop a developer from encroaching in a sensitive wetland habitat. A corporate attorney may represent a particular industrial plant's interest in permit negotiations to set maximum allowable air emission levels for particular pollutants. Another example would be a citizen's group which retains an attorney to file a class action lawsuit against a chemical plant for discharging too much of a toxic substance into a local waterway.

## Educational Preparation

High school students interested in environmental law should take classes which provide a broad educational background in the sciences, government, and communication (including English). Aspiring lawyers must first earn a bachelor's degree as a prerequisite to law school. A good strategy for those dedicated to a career in environmental law would be to earn a bachelor's degree in a technical environmental field (especially a science or engineering) prior to pursuing a legal education.

Environmental lawyers must progress through the traditional legal education process (i.e., three years of law school after receiving a bachelor's degree, followed by successful passage of the bar examination). Law students generally have opportunities to take elective environmental law courses in their second or third year of law school. Many environmental lawyers currently practicing learned their specialty either on-the-job or through postgraduate environmental law courses.

## Potential Employers

**Federal and state government** — Environmental lawyers are employed by both federal and state governments, usually within their environmental protection agencies and justice departments. (The latter are sometimes called "department of the attorney general" or a similar title.)

**Industry** — Large industrial companies employ a limited number of environmental attorneys at the corporate level to track legal developments and to represent the company on environmental matters.

**Law firms** — Environmental attorneys are employed by larger law firms, and by law firms that specialize in environmental matters. In the case of the former, environmental attorneys commonly work on cases in other areas of law (real estate, tax, business contracts, etc.) when there is not enough backlog of environmental work at a firm.

**Environmental interest groups** — Environmental attorneys are employed by national environmental interest groups to bring suits against industrial polluters, government agencies, or other parties to environmental controversies. They also track and recommend legislation and regulations.

### For More Information

Environmental Law Institute
1616 P Street N.W.
Washington, D.C. 20036
(202) 328-5150

American Bar Association
Information Resources Office
750 North Lake Shore Drive
Chicago, IL 60611
(312) 988-5000

## ENVIRONMENTAL JOURNALIST

Every day, environmental stories appear in newspapers, on radio programs, and on television. It is through the news media that the average citizen becomes aware of environmental issues, events, and controversies. The responsibility of bringing reliable, accurate, and timely environmental news to the public belongs to the professional journalist.

### Description of the Work

News reporters and correspondents are often assigned a specialty area of news coverage (commonly referred to as a "beat"), and environmental topics are increasingly identified as one of these areas. Reporters "working the environmental beat" use a variety of means to develop information for their stories. Using press releases submitted by government agencies, environmental interest groups, or industry, they perform additional background research and prepare stories which may have a local, state, national, or international orientation. The actions and opinions of government officials, industrial leaders, and political activists are often closely monitored by environmental journalists, particularly as they relate to controversial issues.

Reporters also investigate leads and news tips which may be called in by citizens. For example, a motorist may report an overturned tanker truck along a highway, which has spilled hazardous chemicals into a local stream. A story such as this is especially attractive to the broadcast news media (television and radio stations) who may be able to report "live" from the scene of the accident. Once the facts of a story have been collected and organized, the reporter determines the particular "slant" or "angle" to take, and writes the story accordingly.

Journalists also strive to present different points of view on an issue to achieve a balanced perspective for the reader. Applying this tenant of journalism to environmental issues, reporters approach many credible sources, within the constraints of time and money, to obtain differing opinions. The most common information sources for environmental stories are regulatory agencies, industries, professors or other impartial experts, environmental groups, and concerned citizens. Environmental reporters obtain information from sources through the use of telephone and face-to-face interview techniques, including on-the-air interviews.

Although at first glance this work appears glamorous and exciting, aspiring environmental journalists also need to be aware of the more "unglamorous realities" of this type of work. Common features of work in the news media are tremendous pressures to meet deadlines; a continually hectic pace; irregular hours to cover accidents, emergencies, or other breaking stories; and no guarantees that a story will not be "bumped" (i.e., not printed or broadcast) by an editor. However, many journalists feel the rewards of having a good story published are more than worth the effort.

## Educational Preparation

Preparing for a career in environmental journalism involves two steps: (1) developing an appropriate educational background, and (2) gaining relevant experience. In terms of educational preparation, there are two choices. Pursuing a bachelor's degree in journalism, with elective classes in the environmental sciences, would provide a good foundation for this type of work. The other approach, equally as good (and maybe better!), is to earn a degree in the environmental sciences, and take elective classes in journalism, English, and other communications subjects.

In addition to an appropriate educational background, gaining journalism work experience is critical to building personal credentials, eventually leading to a professional position. Working on the school newspaper during college is a good start. Most news media organizations, particularly newspapers, offer internship opportunities for prospective reporters. Journalism internships are also offered by trade associations, environmental interest groups, government agencies, and just about any other organization that produces a newsletter, magazine, or other publications. After their education is completed, and some type of experience has been gained, most reporters begin their professional careers at the smaller local newspapers and media organizations where entry-level opportunities are the most plentiful. Later, with some professional journalism experience behind them, reporters can successfully compete for positions with the larger regional, state, and national news organizations.

## Potential Employers

**News services** — News service firms employ journalists to write stories to be distributed to newspapers around the country.

**Newspaper publishers** — The major city newspapers employ journalists to work the "environmental beat" and write news stories and feature articles.

**Television networks** — The national television networks may employ environmental journalists as investigative reporters or as writers for documentaries.

**Radio networks** — The national radio networks may also employ environmental journalists as investigative reporters or as writers for documentaries.

**News magazine publishers** — The major national news magazine publishers employ environmental journalists and photojournalists to prepare regular articles for weekly publication and an occasional special cover or feature story.

### For More Information

Outdoor Writers Association of America
2017 Cato Avenue
Suite 101
State College, PA 16901
(814) 234-1011

National Association of Professional Environmental Communicators
P.O. Box 06 8352
Chicago, IL 60606-8352
(312) 321-3336

## PUBLIC RELATIONS SPECIALIST

Within the larger companies and corporations in America, public relations departments are increasingly looking for specialists in environmental communications. These specialists often have a broad, and sometimes deep, background in environmental issues, as well as the written and verbal communication skills required of the public relations professional.

### Description of the Work

The primary function of a public relations specialist is to assist companies in projecting a positive public image to individual citizens, communities, and the news media. In the environmental arena this can be very challenging as more and more companies find themselves responsible parties at environmentally contaminated sites; the targets of governmental environmental enforcement actions for air or water pollution violations; or attempting to site new incinerator or landfill facilities in areas where the local residents are strongly opposed to such actions (the "NIMBY syndrome" discussed earlier). As a first step in assisting a company in this area, the environmental communications specialist may assist in drafting policy statements on environmental issues, and then work with top management to adopt these statements as official company policy. Once adopted, it is the responsibility of the environmental communications specialist to communicate the policy statements to the public.

Because communication of a company's environmental policies must be followed up by actions to demonstrate a commitment to environmental protection, the environmental communications specialist is becoming more involved in a company's strategic planning process. The decisions made through this process determine how the environmental policies will be reflected in the day-to-day operations of the company.

The public relations or environmental communications specialist implements a wide variety of techniques to build and enhance a company's image as a good, responsible neighbor and citizen. Writing speeches on environmental topics for top managers is a common task. Another frequent activity is the writing and issuing of press releases for both good news (e.g., receiving an environmental award for air pollution control) and bad news (e.g., the company violating its permit limits for toxic chemicals in wastewater). These public relations specialists may also provide managers with training on how to handle the news media, and coach them on effective interview techniques.

In cases where a company is involved in cleaning up a contaminated site, intensive community relations efforts may be required, such as setting up citizen committees to facilitate two-way communications between the company and the affected community. Open houses are sometimes scheduled as a means of getting the public familiar with a manufacturing plant and, perhaps, to display company investments in pollution control equipment. An environmental communications specialist may arrange to have company scientists and engineers available through a speaker's bureau to discuss a company's environmental management program.

Public relations specialists working in the environmental arena also get involved in other aspects of company management. Often they are made responsible for developing a crisis communications plan to respond to environmental accidents and emergencies. They may be requested to assist with a product stewardship program to assist customers with the proper use, handling, and disposal of a hazardous substance. Environmental communication specialists may even get involved in recruiting managers, environmental scientists, engineers, or others for whom the company's environmental commitment is an important factor in accepting an employment offer.

## Educational Preparation

Public relations specialists must have well-developed communication skills, both verbal and written. Those hoping to work on environmental issues should also develop a background in the environmental sciences. Although a specific college degree is not required to become a public relations specialist, students should at least take college level courses in communications, English, writing, and public speaking. In addition, courses in ecology, environmental science, environmental law, and related subjects should be taken to provide sufficient technical background to accurately interpret complex environmental issues for the public.

## Potential Employers

**Government** — Federal and state environmental protection agencies employ public relations specialists in their public affairs offices.

**Industry** — Large manufacturing companies employ environmental public relations specialists at both the corporate and plant levels.

**Environmental interest groups** — These organizations employ public relations specialists primarily to work with the news media in covering environmental stories of interest to the group.

**Public relations firms** — Public relations firms are beginning to employ environmental communications specialists to meet the special needs of their government, industry, and nonprofit organization clients.

## For More Information

Public Relations Society of America
33 Irving Place
3rd Floor
New York, NY 10003
(212) 995-2230

National Association of Professional Environmental Communicators
P.O. Box 06 8352
Chicago, IL 60606-8352
(312) 321-3336

## CONSUMER ADVOCATE

### Description of the Work

Consumer advocates evaluate consumer products available in the marketplace and report the results to manufacturers, government agencies, the news media, or directly to citizens. The author's rationale for including consumer advocates in this discussion is that they are increasingly focusing on the environmental aspects of products.

The decade of the 1990s is witnessing a surge in public awareness about environmental issues, and most citizens want to help protect the environment. The purchase of consumer products that are "environmentally friendly" (i.e., those that do not cause a solid waste problem, those that do not cause significant pollution to manufacture, those that do not waste energy, and those that do not contain toxic chemicals) is an important way for the public to demonstrate its concern.

Environmental consumer specialists coordinate numerous activities to further their cause on behalf of the consuming public. They often become involved in verifying manufacturer's claims of products being "environmentally safe" or "environmentally friendly". This is done through extensive research and product testing efforts. Consumer activists use the results from these product evaluations in a variety of ways, from issuing public statements to filing complaints with manufacturers and/or government regulatory agencies. In approaching manufacturers, a challenge to improve the product in question or to correct product deficiencies is one strategy used by consumerists. Other activities of the consumer advocate include: conducting satisfaction surveys of product users and reporting the results; testifying at public hearings and before legislative committees; and waging public information campaigns for more and improved "environmentally friendly" products (efforts usually targeted at manufacturers, as well as politicians and government regulatory agencies).

Consumer advocates might be employed by public interest groups, print and broadcast news media organizations, and publishers of consumer-oriented magazines. In addition, many of these professionals work as freelance consultants or writers.

The chances of achieving success as an environmental consumer advocate are greatly improved with the development of personal communication and organization skills, and a broad environmental background. Exceptional writing and public speaking abilities will help the consumer advocate effectively construct and transmit clear and concise messages. Also, a familiarity with political and regulatory processes allows the consumerist to focus time and effort where the greatest impact can be made. Finally, some educational background in the environmental sciences, and some familiarity with different types of industrial manufacturing processes will provide a solid foundation from which the consumer advocate may evaluate the environmental costs and liabilities of products in the marketplace.

## Educational Preparation

Although no special educational background is mandatory to become a consumer advocate, there are certain college level courses which will provide a sound technical background from which to evaluate consumer products. Such courses include: chemistry, physics, mathematics, environmental engineering, and ecology. A bachelor's degree is the minimum educational requirement for professionals working in this field.

## Potential Employers

**Government consumer agencies** — The federal and most state governments have an agency, staffed by consumer advocates, dedicated to the protection of the consumer.

**Consumer service firms** — Independent consumer product evaluation companies also employ consumer advocates to evaluate new products.

**Environmental interest groups** — As "green" and "environmentally friendly" labels are becoming popular with consumer product manufacturers, environmental interest groups are beginning to employ consumer advocates to evaluate and validate such claims.

### For More Information

American Council on Consumer Interests
240 Stanley Hall
University of Missouri
Columbia, MO 65211
(314) 882-3817

Consumer Federation of America
1424 16th Street N.W.
Washington, D.C. 20036
(202) 387-6121

Society of Consumer Affairs Professionals in Business
4900 Leesburg Pike
Suite 400
Alexandria, VA 22302
(703) 998-7371

## DATA PROCESSING SPECIALIST

### Description of the Work

A substantial challenge facing both government and industry in the 1990s is the efficient gathering, organizing, analysis, and file storage of large volumes of technical environmental information. The key tool in addressing this need is the computer. Some type of data processing is involved in virtually every aspect of technical environmental work. Over the last decade, data processing work in the environmental field was performed either by computer specialists or by environmental scientists or engineers. Both groups had their limitations, as the computer specialists did not fully understand the data processing needs of environmental professionals, and the environmental scientists and engineers did not fully comprehend the technical complexities of data processing. A new professional niche is developing for computer specialists with training in environmental data processing applications, particularly for the development and operation of large-scale data bases.

Data processing specialists generally focus on one of two areas of emphasis, systems analysis or programming. Systems analysts develop and implement the overall plans for new computer systems, or enhancements to existing systems. They are responsible for understanding complex data processing tasks, then integrating computer hardware and software (programmed instructions) into a system to effectively and efficiently perform those tasks. In designing a data processing system, the systems analyst must determine all of the functional requirements, from how data needs to be entered into the system, to how various computer elements will communicate with each other.

The programmer, on the other hand, writes detailed logical instructions, referred to as programs, for the computer to follow. These programs are written for environmental applications with input and assistance of environmental scientists and engineers. Once written, the program is tested by using sample data. The programmer then corrects all errors in the program until the program runs properly. Programs for environmental applications can become very complex, such as computer models which depict the potential movements of contaminants through the ground water, or projects how certain air pollutants will disperse under differing weather conditions. Lastly, computer programmers write specific operating instructions for program users, and often provide consultation services to users when problems arise.

## Educational Preparation

Data processing specialists usually hold a minimum of a bachelor's degree in computer science. Those interested in data processing in the environmental field should develop technical familiarity with various environmental disciplines by taking elective classes in these areas.

## Potential Employers

**Government** — Both federal and state environmental protection agencies employ data processing specialists to assist in managing extremely large volumes of data involved with regulatory programs.

**Industry** — Manufacturing industries employ data processing specialists to develop data bases for environmental information needed to maintain compliance with laws and regulations, and to control inventories and usage of toxic and hazardous chemicals.

**Data processing consulting firms** — Data processing specialists are also employed by consulting firms to assist government and industrial clients in building and managing large data processing systems.

### For More Information

National Society for Computer Applications in Engineering, Planning, and Architecture
5 Park Avenue
Gaithersburg, MD 20877-2915
(301) 926-7070

Data Processing Management Association
505 Busse Highway
Park Ridge, IL 60068
(312) 693-5070

## PHOTOGRAPHER

Photography is playing an increasingly important role in the communication of environmental information. Whether as artists or as technicians, photographers are finding new opportunities to specialize with environmental subjects.

### Description of the Work

Technical photographers working in the environmental field strive to accurately record and document specific places, events, techniques, conditions, and other situations where important information is derived or interpreted from photographic images. Examples of assignments might include: aerial photography of a site being considered for a landfill; photographing evidence of hazardous waste violations in support of governmental enforcement actions; photography of new environmental technologies for publication in books and magazines; photography to document "before and after" results of experiments, such as microbes breaking down toxic contaminants in the soil; or photography of an oil spill, or other newsworthy events, as a photojournalist.

Photographers working in an artistic mode, on the other hand, attempt to capture or portray more abstract ideas, such as feelings, personalities, moods, gracefulness, power, energy, etc. To enhance an image's artistic aspects, photographers use an entire spectrum of techniques which create special effects. Creatively composed photographic images of wildlife, forests, mountains, rivers, landscapes, seascapes, cities, industry, pollution, and other related subjects can transmit emotionally powerful messages to the viewer.

## Educational Preparation

Although there are no specific educational requirements for becoming a photographer, most professional photographers obtain substantial technical training through special schools, apprenticeships, seminars, and/or on-the-job training. Special training requirements do exist for some photographers, such as those who gather photographic evidence for court cases.

## Potential Employers

**Government** — Photographers are employed by federal and state environmental protection agencies to gather evidence in enforcement cases, and occasionally to work on public relations assignments.

**Publishers** — Newspaper and magazine publishers employ creative photographers and photojournalists to provide visual images to enhance a story, or simply to display creative photographic artwork.

## For More Information

Professional Photographers of America
1090 Executive Way
Des Plaines, IL 60018
(708) 299-2685

Evidence Photographers International Council
600 Main Street
Honesdale, PA 18431-0351
(717) 253-5450

# ARTIST

## Description of the Work

Artists, using music, paint, sculpture, political cartoons, poetry, creative literature, folk crafts, and other media, can make significant contributions to raising public awareness about the environment. Whether an artistic work reflects the tranquility, violence, or beauty of the natural world, or relates the adverse impacts of mankind on the integrity of the environment, these messages can be effectively communicated to the public through works of art. Because of the potential significance artwork can have on an individual admirer, acknowledgement of artists as environmentalists warrants mention here.

## Educational Preparation

No special educational requirements exist for artists, although most successful professional artists have had substantial college level training.

## Potential Employers

Artists are often self-employed. They may both create and sell their work on a freelance basis. Many artists work to get commissioned by a business, government agency, or foundation to create special works of art. Other artists may find employment in special niche areas, such as technical illustration for textbooks.

CHAPTER 7

# Careers in Natural Resources Management

# Careers in Natural Resources Management

This category of career opportunities covers the wide range of professions that involve the planning, management, and wise use of natural resources, such as land, forests, wildlife, parks, water, fisheries, and the like. This chapter focuses on what are sometimes referred to as the "traditional" natural resource disciplines. Most of these career opportunities are in some way involved with how federal, state, and local governments manage the use of resources for the benefit of the public. However, there are also opportunities with private large-scale landowners and nonprofit public interest groups.

Individuals who choose careers in the natural resources management area tend to be more dedicated to stewardship of land, water, and wildlife than to maximizing their financial position. The pay for these jobs tends to be moderate overall, but the professional satisfaction from contributing to the betterment of the planet can be great. Because of this job satisfaction, natural resource management professionals are more likely to remain with a single employer for long periods of time, as opposed to other environmental professionals. The author has known many "lifers" in the fields of forestry, parks, and wildlife and fisheries biology.

Careers in natural resources management generally offer more opportunities to work outdoors than some of the more purely scientific, engineering, or legal types of environmental careers. The number of available positions in these careers is limited, however, and individuals may need to be patient and persistent with potential employers in order to eventually get hired.

One growing area of work within the natural resource management professions is the assessment of environmental impact from new development and pollution on natural resources. By understanding how certain actions may adversely affect land, fish and wildlife, water, forests, and other resources, project planners can modify their plans to minimize such impacts. Regulations at both state and federal government levels require that such assessments be performed prior to the issuance of public funds for a major project. Also, with privately funded projects, often such assessments are performed to address concerns raised by the public or by agencies issuing permits for a project.

## PARK AND RECREATION PROFESSIONAL

### Description of the Work

Outdoor recreation professionals are responsible for the planning, development, and management of a wide variety of outdoor recreation resources, from isolated wilderness areas to intensively used urban parks. Because several of these careers are closely related, they are grouped together in this discussion: park ranger, park manager, and recreation planner. Another related career is that of the interpretive naturalist, which is covered in the chapter on environmental education.

**Park ranger** — Park rangers perform a wide variety of tasks related to managing a park's natural resources and facilities, and enforcing the established rules and regulations. During a single week, a ranger may get involved with such diverse activities as taking user fees at an entrance gate, patrolling a campground, issuing backcountry permits to backpackers, removing a problem animal from a picnic area, rescuing a stranded boater, talking to a group of school children about having respect for nature, checking fishermen for proper licenses, supervising a staff to set up a stage for a special event, investigating an incident of vandalism, or monitoring the activities of a park concessionaire.

**Park manager** — Park and recreation facility managers are responsible for all aspects of day-to-day park operations, including administrative and community relations functions. Included in this discussion of park managers are the managers of related outdoor recreation facilities, such as docks and boat ramps, fishing access areas, campgrounds, etc. Park management professionals must balance the intensity of public recreational use of an area with the need to manage and protect its natural resources. Success at achieving such a balance often requires a broad set of personal management skills which include communications, budgeting, policy making, staff supervision, human relations, and organizational skills. Such skills are put to use by the park manager on a daily basis in negotiating with an agency's top management to increase the annual budget; hiring and firing laborers who perform park maintenance functions; establishing a good working relationship with a park concessionaire; developing and implementing programs to improve public and employee safety at a park. The diversity of work activities for park managers is extremely broad, and may include: overseeing construction activities for a new restroom or beach house facility; giving a speech at a local civic organization luncheon; working with the local health department to investigate a water quality problem; forging a cooperative arrangement with the sheriff's department for emergency medical evacuations from the park; inspecting an area of a park damaged by off-road vehicles; dealing with an angry camper whose campsite was flooded during a storm; and establishing goals and objectives for future park improvements.

**Recreation planner** — The outdoor recreation planner is concerned with the planning, design, and development of park and recreation resources to meet the needs of all segments of society. The planning process begins with the collection and analysis of information on the public demand for recreational services and facilities. After identifying the areas of greatest need, recreation planners establish goals and objectives for new recreation development. Specific land areas and natural resources are identified and analyzed to determine their suitability for development. Once a site is selected, land acquisition and conceptual design begins. Final design, financing, and construction follow. Planners must cultivate citizen participation throughout the planning process to assure that the established objectives are met.

On any given day, a recreation planner might be involved in any of the following activities: preparing a draft report on user conflicts between canoeists and trout fishermen on a stream within a proposed park site; meeting with the local chamber of commerce to garner support for a specific project; presenting a proposal to a citizens group regarding the renovation of a dilapidated fishing pier in an urban area; preparing a preliminary park design for a landscape architect consultant; conducting a statistical analysis of the results of a local recreation survey; negotiating with landowners located adjacent to a natural area to create a vegetative buffer along the park boundary; developing a strategy for land acquisition; reviewing an ecological analysis of a potential park site; or scheduling construction activities.

## Educational Preparation

**Park ranger** — To effectively perform a wide range of tasks, park rangers must enjoy working with people, have a sound knowledge of conservation laws and regulations, and have some background in natural resource management. People who have become park rangers over the years have had very different types of educational preparation, from individuals with degrees in teaching, sociology, law enforcement, and biology to those with only high school, vocational, or military training. Because of the high level of interest for a relatively limited number of positions, park rangers are being required to have at least a bachelor's degree. Increasing problems with drugs, alcohol, violence and vandalism in public parks are making those job candidates with law enforcement backgrounds particularly attractive to government employers. In addition to law enforcement training, college level courses in the natural sciences (i.e., ecology, biology, botany, and geology), the social sciences (especially psychology and sociology), and natural resources management (i.e., recreation management, forestry, fisheries, range management, etc.) could be helpful to the prospective park ranger.

**Park manager** — As with park rangers, recreation management professionals have come from a variety of educational backgrounds. Today, however, many colleges and universities offer specific programs in parks and recreation management. In addition to, or as part of, a parks and recreation core curriculum, the following college level subjects may prove useful to future candidates: psychology, sociology, political science, law enforcement, ecology, economics, conservation law, landscape architecture, and natural resources management (i.e., forestry, fisheries, land use planning, etc.). A bachelor's degree for entry-level recreation professionals has become a standard.

**Recreation planner** — Educational preparation for a person aspiring to become a recreation planner should include college courses in most of these subjects: sociology, landscape architecture, urban and regional planning, economics, political science, ecology, natural resource management, park management, statistics, and planning/zoning law.

## Potential Employers

**Federal government** — Federal agencies that employ park and recreation professionals include: the National Park Service, the U.S. Forest Service, the Fish and Wildlife Service, the U.S. Army Corps of Engineers, the Bureau of Land Management, the Bureau of Reclamation, and the Bureau of Outdoor Recreation.

**State government** — Most states employ park and recreation professionals with their natural resources or fish and game departments, where they are involved in managing state parks, state forests, state natural areas or wilderness areas, and programs providing financial support to local recreation agencies.

**Local government** — Municipal, county, and regional or metropolitan agencies regularly employ parks and recreation professionals to manage local parks, recreation centers, and related facilities.

**Utilities companies** — Large commercial utility companies around the country often own and manage significant acreages of land, usually encompassing impoundments, hydroelectric plants, coal fired power plants, or nuclear power generating stations. As a public service, such companies operate picnic areas, boat launches, bathing beaches, campgrounds, and interpretive centers. These facilities are usually managed by park and recreation specialists.

**Timber companies** — Large paper and timber companies own vast forest lands, sometimes several square miles in size. Most of these companies allow for hiking, camping, fishing, boating, and hunting activities by the general public. Recreation managers are employed by these companies to operate and maintain campgrounds, public access sites, trails, picnic areas, and other facilities.

## For More Information

National Recreation and Park Association
3101 Park Center Drive
Alexandria, VA 22302
(703) 820-4940

# WILDLIFE BIOLOGIST

## Description of the Work

Wildlife specialists work to manage wild populations of animals through habitat management, introductions and reintroductions of species, review of environmental impacts of development projects, health evaluations and disease control, and other techniques. This section discusses two types of wildlife careers, the wildlife biologist and the game area manager.

Wildlife biologists are concerned with the protection and long-term management of wildlife species, usually for one of two distinct purposes. The first is for the purpose of game management, to support some level of recreational hunting activity. The other purpose is for the protection or preservation of certain wildlife species to maintain the ecological diversity and integrity of the natural environment.

For game management purposes, wildlife biologists may get involved in some of the following activities: assessing the quality of white-tail deer habitat in a local area to assure the survival of juveniles through a particularly harsh winter; evaluating the advantages and disadvantages of introducing an Asian variety of game pheasant to replace the declining ring-necked pheasant populations; studying the impacts of a disease on a natural elk herd population; evaluating the toxic affects of lead bird shot on certain waterfowl species; appearing as an expert witness at a public hearing on government policies regarding the trapping of fur-bearing animals; or determining the maximum allowable number of specific animals to be taken during hunting season. In addition to animal-oriented tasks, game area managers must also focus on managing the wildlife habitats within their property boundaries. These activities often include: managing the planting and harvesting of food crops or browse to support wildlife; hatching and raising game bird species for supplementing natural populations; and conducting administrative activities such as budgeting, staff supervision, equipment and facility maintenance and repair, site security, and related functions.

For the purposes of protection and preservation of individual wildlife species, wildlife biologists have a slightly different set of duties. Some of the following types of activities may be included: studying the ecological requirements of legally protected, threatened and endangered species; identifying natural habitat areas for public acquisition as wildlife habitat; assessing the quality of a forest habitat to support a protected owl; constructing nesting platforms in a marsh for use by osprey; removing a bear from a campsite or a racoon from a residential area; researching the population dynamics of pest species of field mouse; promoting a program to raise revenues for non-game wildlife management in a particular state; monitoring predator-prey relationships between wolves and moose at a national park; or analyzing the potential negative affects of construction of an oil pipeline on the migratory patterns of caribou.

Wildlife biologists are currently working to address several difficult and complex resource management issues. Examples of these issues are the continued destruction of critical wildlife habitat, particularly from urban development; the maintainence of political support and funding for the protection of threatened and endangered species; the reintroduction of species to their original, restored habitat; the evaluation of proposals by hunting organizations to introduce exotic species into an area solely for game sport; and the determination of the effects of environmental degradation and toxic chemicals on wildlife species.

## Educational Preparation

Students wishing to become wildlife biologists should begin their training in high school by taking available classes in biology, chemistry, and mathematics. College coursework should include continuing classes in biology, chemistry, and math along with physics, ecology, and wildlife management. Additional courses in economics, sociology, and political science will also prove helpful.

## Potential Employers

**Federal government** — The U.S. government employs wildlife biologists in a number of agencies, including: the Fish and Wildlife Service, the U.S. Forest Service, the National Park Service, the U.S.

Army Corps of Engineers, the Bureau of Land Management, and the Bureau of Reclamation.

**State government** — State governments employ wildlife biologists in their fish and game departments or natural resources departments where they manage wildlife in game areas, state parks, state forests, and in urban areas.

**Industry** — Companies which own and manage large tracts of land, such as paper and timber firms and electric utility companies, occasionally employ wildlife biologists to manage habitats, game populations, and hunting activities.

**Environmental organizations** — Several nonprofit environmental organizations employ wildlife biologists to manage private natural areas or to advise on wildlife issues. Examples of such organizations include: the Nature Conservancy, the National Audubon Society, the National Wildlife Federation, the Sierra Club, as well as a host of regional/state/local nature organizations.

### For More Information

The Wildlife Society
5410 Grosvenor Lane
Bethesda, MD 20814-2197
(301) 897-9770

Wildlife Management Institute
1101 14th Street N.W.
Suite 725
Washington, D.C. 20005
(202) 371-1808

## FISHERIES BIOLOGIST

### Description of the Work

The fisheries biologist is primarily responsible for the management and protection of aquatic life. The goals of fisheries management may include: the sustaining of naturally reproducing fish species for commercial or recreational uses, the planting of fish, the introduction of certain exotic species to address a specific problem, the reintroduction of aquatic species in places where they have disappeared (to re-establish the ecological integrity of a resource), or the protection of threatened and endangered species. Fisheries biologists often become involved in one or more of the following activities: conducting spawning studies of key species (such as large mouth bass) in an inland lake; studying a particular disease that is reducing a population of forage fish (such as alewives on the Great Lakes); determining the maximum allowable take of various species under a single fishing license; developing programs to manage pest species (such as the sea lamprey in the Great Lakes that was adversely affecting lake trout populations); evaluating the introduction of an exotic species, like grass carp, to manage weed growth in a small lake; reintroducing a formerly native species of trout in a reclaimed trout stream (such as the grayling on the AuSable River in Michigan); proposing special fishing regulations for sensitive stretches of a high quality trout stream; monitoring the harvest of fish by commercial fishermen; developing policy recommendations on Indian fishing rights on former territorial waters; identifying potential sites for public fishing access to rivers, lakes, and the ocean; determining the impacts of shoreline or coastal developments on fish habitat; or developing improved commercial fishing techniques to minimize the kill of non-commercial species. Some of the critical challenges facing fisheries biologists today are how to protect and improve fish spawning habitats, clarifying Indian fishing rights in treaty waters, and managing fisheries in areas of degraded water quality.

Other professions related to the fisheries biologist are the fish hatchery manager and the limnologist. Fish hatchery managers are responsible for the rearing of fish for planting programs. In operating a hatchery, managers may be concerned with: the artificial water conditions for spawning and rearing of fish (e.g., temperature, flow velocity, etc.); conducting growth and health studies; determining the proper time for releases; and development of efficient feeding programs. The limnologist is concerned with the ecological quality of the whole freshwater stream environment, including fish, benthic organisms, and aquatic plants.

## Educational Preparation

High school students interested in fisheries biology should take a strong academic program in the sciences, stressing courses in biology, chemistry, and mathematics. The college level fishery management curriculum is often quite rigid with respect to requiring specific fishery management and biology classes. Good choices for elective classes would include physics, hydrology, water quality management, economics, political science, and sociology.

## Potential Employers

**Federal government** — The majority of fisheries biologists employed by the federal government work for the U.S. Fish and Wildlife Service, who often provide service to other federal agencies. Some fisheries biologists are also employed by the Bureau of Land Management, the National Park Service, and the U.S. Forest Service.

**State government** — States employ fisheries biologists in their departments of natural resources, conservation, or fish and game management.

**Industry** — A relatively new employer of fisheries biologists is the aquaculture (or fish farming) industry.

**Research institutes** — Usually associated with universities, industries, or nonprofit agencies, research institutes occasionally employ fisheries biologists to carry out scientific experiments involving fish and other aquatic species.

**Environmental groups** — Nonprofit environmental and conservation organizations also employ fisheries biologists to conduct studies and advise on resource management issues.

## For More Information

American Fisheries Society
5410 Grosvenor Lane
Suite 110
Bethesda, MD 20814-2199
(301) 897-8616

# FORESTER

## Description of the Work

Foresters and forestry specialists are responsible for the protection and wise use of trees and related forest resources. They are trained to perform a full spectrum of forest management tasks, including: estimating usable standing timber in a forest tract; fighting and managing forest fires; determining optimal planting-harvest rotations; protecting the hydrological patterns of a forest during harvesting; identifying and managing tree diseases; protecting the natural integrity of a forest ecosystem; managing recreational use of forests; assessing the damage to a forest caused by acid rain; preventing erosion on forest lands; managing forests as optimal habitats for various species of wildlife (game, non-game, and threatened/endangered species); developing efficient logging techniques that minimize environmental damage; planning the locations of access roads for timber operations; identifying forest areas to be protected as wilderness; and educating private woodland owners on sound forest management techniques.

Some foresters specialize in specific aspects of forest resource management. Examples include the urban forester, the forest pathologist, and the forest fire specialist. The urban forester focuses on the special problems of planting and maintaining healthy trees in a city environment. These problems might include: identifying specific tree species that are tolerant of road salt, compacted soils, or air pollution; coordinating the cleanup of debris after a wind storm; overseeing the care and maintenance of landscape plantings on public property. To the urban forester, the "forests" might be along road right-of-ways, in city parks, or on public school property. A different type of forestry specialty is the forest pathologist. This specialist focuses on conducting research on the causes and dynamics of tree diseases, and develops techniques to prevent the spread of such diseases. The key areas of study for the forest fire specialist are fire science and fire ecology. This person understands the role of fire in the natural ecosystem, the behavior of fire in the forest, and strategies and techniques for effectively managing controlled burns as well as uncontrolled wild fires.

## Educational Preparation

High school students interested in forestry should take a strong academic program in the sciences, stressing courses in biology, chemistry, and mathematics. The college level fishery management curriculum is often quite rigid with respect to requiring specific forestry management and biology classes, such as woody plant identification, forest soils, forest hydrology, forestry economics, fire ecology, forest pathology, forest ecology, and timber harvesting. Good choices for elective classes would include physics, hydrology, water quality management, economics, political science, and sociology.

## Potential Employers

**Federal government** — Most foresters employed by the federal government work for the U.S. Forest Service. Other federal agencies which employ foresters include the National Park Service, the Bureau of Land Management, and occasionally the Department of Defense.

**State government** — States employ foresters in their departments of natural resources or conservation.

**Local government** — Growing numbers of foresters are finding employment with municipalities to manage urban forestry programs.

**Industry** — Paper and timber companies employ foresters to manage timber planting and harvesting programs. Occasionally, other companies owning large tracts of land, such as utility companies, may also employ foresters.

## For More Information

Society of American Foresters
5400 Grosvenor Lane
Bethesda, MD 20814
(301) 897-8720

The American Forestry Association
1516 P Street N.W.
Washington, D.C. 20005
(202) 667-3300

# RESOURCE GEOLOGIST

## Description of the Work

Geologists study the origins, history, structure, and dynamic processes of the earth, and apply this knowledge to solve practical problems. Many geologists do not work in the field of resource management, but rather are involved with the exploration of economically recoverable mineral deposits, and with the extraction of mineral, oil, and gas resources. This section will address only the work of geologists that directly relates to natural resource management, protection, and reclamation.

Many geologists are employed by resource management agencies to conduct inventories of geological resources. They gather data through research activities, field testing, and surveying. The geological information is then plotted geographically to create inventory maps. Sometimes such inventory activities focus on special problems (e.g., identifying and mapping areas of geological hazards, such as areas vulnerable to earthquakes, mudslides, and coastal erosion; or conducting a survey to identify and map unique geological features and formations which need to be preserved and protected for research and educational purposes).

Another type of work performed by government geologists is the development and enforcement of policies and regulations. Much of this work is oriented towards the mining, petroleum, and other resource extraction industries. Federal and state regulations address issues such as the quantity of the resource being mined, how the material is being extracted (i.e., in a safe and environmentally sound manner), and how wastes are handled. Other regulations mandate that mining excavations be "reclaimed" (usually involving regrading and revegetating an area) so that the land can be used for other purposes once the mining operations have ceased. Some states and local governments have special geological regulations to address specific conditions, including controlling new development on steep hillsides, eroding shorelines, along geological faults, and in ecologically sensitive sand dune areas. Geologists working as regulators conduct inspections, issue permits and citations, work out agreements with regulated companies and landowners, and occasionally appear in court as part of enforcement proceedings.

Some geologists specialize in the study and management of ground water resources. In many parts of the country, ground water is a critical source of water supply for both municipalities and farmers. Geologists, or hydrogeologists as they are sometimes called, are involved in estimating and monitoring the quantity of ground water in complex underground formations called aquifers. If aquifers are pumped too rapidly, the entire ground water system may become vulnerable to contamination. Aquifers are replenished from ground surfaces, termed "recharge zones". If too much development occurs over these recharge zones, the aquifer will not receive enough water to be recharged and may eventually become unusable. Therefore, geologists also assist in the development of regulations for controlling the rate and allocation of ground water withdrawals from specific aquifers, as well as controlling development in recharge zones.

## Educational Preparation

High school students interested in geology should develop a strong science (including biology and chemistry) and mathematical background. At the college level, geology core courses are usually specified. Elective courses in physics, soil science, hydrology, economics, and political science will also prove useful to a geologist.

## Potential Employers

**Federal government** — Most resource geologists employed by the federal government work for the various agencies of the Department of the Interior, including the Bureau of Mines and the National Park Service.

**State government** — State agencies employ resource geologists in their natural resources or conservation agencies, usually to regulate mineral extraction activities or to develop state geological surveys.

**Industry** — In private industry, resource geologists are employed by mining, exploration, and oil and gas development companies.

## For More Information

American Geological Institute
4220 King Street
Alexandria, VA 22302-1507
(703) 379-2480

Geological Society of America
3300 Penrose Place
P.O. Box 9140
Boulder, CO 80301
(303) 447-2020

National Water Well Association
6375 Riverside Drive
Dublin, OH 43017
(614) 761-1711

Association of Ground Water Scientists and Engineers
6375 Riverside Drive
Dublin, OH 43017
(614) 761-1711

American Association of Petroleum Geologists
P.O. Box 979
Tulsa, OK 74101
(918) 584-2555

Association of Engineering Geologists
323 Boston Post Road
Suite 2D
Sudbury, MA 01776

# URBAN AND REGIONAL PLANNER

## Description of the Work

Urban and regional planners are concerned with the environmentally compatible growth and development of communities and regions. These professionals often specialize in one or more specific aspects of planning, such as land use, water resources, transportation, waste management, housing, industrial and commercial development, public facilities, parks and open space, and medical facilities. Planners are usually employed by government agencies or by consulting firms serving government agencies.

The planning process involves a wide variety of duties that ultimately lead to the preparation, presentation, and implementation of community development plans that maximize benefits to the public. Examples of tasks that may be part of a planning process include: gathering and analyzing information about a city or region; conducting public opinion surveys to assist in developing goals and objectives; developing alternative community development scenarios; analyzing alternatives in light of community goals and objectives; selecting and recommending a best alternative and presenting it in a plan to a public body for adoption; and developing tools for implementing a plan (e.g., local regulatory ordinances, funding mechanisms, site plan review processes).

The types of urban and regional planners that most often are involved in managing and protecting the environment are those that specialize in land use, water quality, and solid waste management.

Land use planners are primarily concerned with assuring that new land development projects are compatible with existing surrounding land uses. For example, a land use planner might propose a zoning ordinance change so that an industrial plant would not be located immediately adjacent to a residential neighborhood. Other examples would be establishing a program to protect prime agricultural land near a city from residential construction, or preparing a plan to guide commercial development away from environmentally sensitive floodplain areas.

Water quality planners make sure that new growth and development in a community does not further pollute local streams, rivers, and lakes. They assure that existing or planned public sewage treatment systems have adequate capacity to accommodate new development projects. Storm water management is another concern of the water quality planner. Rainfall runoff from roofs, parking lots, city streets, industrial plants, and construction sites contribute substantial pollution to local waterways if not properly managed. A planner might develop rules to require a land developer to construct storm water basins as part of a shopping mall project to allow for pollutants to settle out of parking lot runoff water before being discharged into a stream.

Environmental planners work to find environmentally suitable sites for various types of large-scale development projects (such as locating a site for a new pulp and paper mill or chemical plant), and to minimize the environmental degradation caused by such projects. In cases where there may be public opposition to a project, such as a proposed landfill or incinerator, planners sometimes get involved in community relations and participation activities. Planners often coordinate teams of environmental specialists (geologists, ecologists, engineers, etc.) who collect and analyze data on soils, bedrock geology, surface drainage, vegetation, site drainage and topography, surrounding land uses, ground water, wildlife and fisheries, and other site information. Using overlay analysis techniques and other methods, environmental planners identify and plot those areas within a proposed project site that are unsuitable for development, as well as those that are the most suitable.

Planners also advise development teams (usually consisting of architects and engineers) on ways to minimize the environmental impact of a project. Some of these methods include: using a natural area buffer between an industrial building and a public park to filter out noise and dust; construction of storm water retention basins to hold and settle out pollutants from surface runoff; and planting vegetation in drainage ditches and along waterways to prevent erosion and sedimentation.

## Educational Preparation

High school students considering a career in urban and regional planning should develop a background in the sciences, mathematics, civics, and social studies. At the college level, the urban and regional planning curriculum is usually structured with core and elective courses. Aspiring planners desiring to specialize in environmental planning should take elective courses in ecology, remote sensing, water quality management, environmental policy and law, and environmental engineering.

## Potential Employers

**Federal government** — The federal government employs urban and regional planners in a wide variety of agencies, although the position titles may be related to program or administrative functions rather than "planner". These agencies include: the Environmental Protection Agency, the National Park Service, the U.S. Forest Service, the Bureau of Land Management, the National Oceanic and Atmospheric Administration, the Department of Housing and Urban Development, the U.S. Army Corps of Engineers, the Department of Health and Human Services, and the Department of Transportation. In addition, federally authorized agencies, such as the Tennessee Valley Authority, employ urban and regional planners.

**State government** — States also hire urban and regional planners in agencies involved with environmental protection, public health, housing, transportation, commerce, and agriculture.

**Sub-state regional and metropolitan planning agencies** — These multi-city and multi-county agencies employ urban and regional planners to analyze and plan for growth and development in specific geographic areas.

**Local government** — Individual cities, townships, and counties employ urban and regional planners to manage and carry out local planning, zoning, and community development programs.

**Consulting firms** — Planning, landscape architecture, architecture, and engineering consulting firms employ urban and regional planners to assist on large development projects, as well as to offer technical planning assistance to communities and businesses.

**Other potential employers** — Urban and regional planners are also employed by land developers, banks, and public utility companies.

## For More Information

American Planning Association
1776 Massachusetts Avenue N.W.
Washington, D.C. 20036-1997
(202) 872-0611

# LANDSCAPE ARCHITECT

## Description of the Work

Landscape architects are involved with the design, planning, and development of the land, to enhance the quality of life and to solve specific man-environment problems. Projects range in scale from small urban "pocket parks" to national forests consisting of hundreds of square miles of land. Their work may involve: the design of office parks and shopping malls, golf courses, zoos, marinas, urban plazas, and open spaces; or redevelopment of historical areas, commercial districts, or abandoned industrial or mine sites. Landscape architects might also prepare master plans for recreational facilities such as trail systems, highway rest areas, campgrounds, or parkways; or inventory and analyze the natural resources of a site.

Working at the interface of people and the environment, landscape architects prepare designs that: preserve and protect the integrity of natural systems; conserve and protect areas with historical character and environmental sensitivity; blend land development projects into the natural landscape; create safe, usable, and comfortable outdoor spaces and corridors; and use natural features and materials to control light, shade, temperature, drainage patterns, wind, pedestrian and vehicular movement, and other conditions associated with a particular site.

There are approximately 25,000 landscape architects currently practicing in the U.S. These professionals work in private landscape architecture firms, land development companies, academia, and federal, state, and local government agencies.

Prospects for recent and future landscape architecture graduates are bright as the Bureau of Labor Statistics (U.S. Department of Labor) has identified landscape architecture as one of the growth professions of the 1990s.

## Educational Preparation

A bachelor's degree in landscape architecture is generally considered the minimum requirement for entering the profession. Many practicing professionals have earned master's degrees as well.

There are over 40 accredited college and university programs in landscape architecture throughout the country. The curriculums include courses in natural sciences, social sciences, mathematics, art and aesthetics, history of landscape architecture, plant materials, site analysis, construction techniques, computer applications, office practice, design methods and techniques, and design studios.

In addition to a professional degree, most states require that landscape architects be licensed to protect the public's health, safety, and welfare. Licensing criteria often require the applicant to hold a degree in landscape architecture, pass a national licensing examination, and complete a period of supervised practice.

To prepare for entry into a collegiate landscape architecture program, high school and undergraduate students may find the following types of courses useful: biology, botany, art/design, mathematics, psychology, sociology, public speaking, and composition.

## Potential Employers

**Federal government** — The federal government employs landscape architects to design and oversee the development of outdoor recreation facilities. Most landscape architects employed by the federal government work for the National Park Service and the U.S. Forest Service.

**State government** — State departments of natural resources or conservation also employ landscape architects, primarily for the design and development of recreation areas and facilities.

**Consulting firms** — A large number of landscape architects work for landscape architecture, planning, architecture, and engineering consulting firms where they provide design services for industrial parks, apartment complexes, college campuses, commercial developments, and homeowners, as well as for government clients.

### For More Information

American Society of Landscape Architects
4401 Connecticut Avenue N.W.
Fifth Floor
Washington, D.C. 20008-2302
(202) 686-ASLA

## REFERENCES

*Between People and Nature: The Profession of Landscape Architecture,* brochure by American Society of Landscape Architects.
*Landscape Architecture: Shaping Our Land,* brochure by American Society of Landscape Architects.

## REMOTE SENSING SPECIALIST

### Description of the Work

Remote sensing specialists and related professionals are concerned with gathering, analyzing, interpreting, and displaying natural resource information for use by scientists and policy-makers. This information is developed from remote sensing techniques, such as aerial photography, color infrared photography, radar imagery, and multispectral scanning (satellite imagery). Computer-assisted design and automated geographic information systems are often critical tools for the remote sensing specialist. Using these technologies, complex natural resource and environmental problems can be analyzed from an entirely different perspective than can be seen from the ground.

Activities that a remote sensing specialist may perform include the following: interpreting aerial photographs or satellite images; measuring various landscape features; classifying land uses and vegetative cover types for government planning purposes; counting the number of healthy and diseased deer in a densely forested area using infrared techniques; delineating a particular water pollution plume from a mining operation; estimating the area of tropical rain forest lost to slash and burn agricultural activities; determining the extent of a deadly tree disease within a national forest; classifying the major wetland types within the proposed boundaries of an airport expansion project; measuring the extent of urban sprawl over prime agricultural lands within a metropolitan area; evaluating the hydrologic patterns of an area degraded by surface mining; developing a data overlay program to analyze land use changes over a ten-year period; or assessing the impact of acid rain on forests in the Appalachian Mountains.

It is common for remote sensing specialists to have a second scientific area of expertise, such as geology, wildlife management, forestry, water resources development, or land use planning. Having a second area of specialization allows this professional to more fully apply remote sensing technologies to particular problems, and to more completely interpret remotely sensed data.

## Educational Preparation

High school students interested in remote sensing can get a head start on college preparation by taking courses in biology, chemistry, mathematics, and computer operation. Although few universities offer degrees in remote sensing per se, the following courses will prepare students well for such work, or for further graduate study: geography, cartography, remote sensing techniques, physics, ecology, biology, geology, computer sciences, and statistics. In addition, students should develop a specialty area of remote sensing application, such as forestry or water resources management.

## Potential Employers

**Federal government** — Remote sensing specialists are employed by federal agencies that either manage large land areas or are involved in scientific research activities. Examples of these agencies are the National Park Service, the U.S. Forest Service, the Bureau of Land Management, the Environmental Protection Agency, the U.S. Geological Survey, the National Oceanic and Atmospheric Administration, the National Aeronautics and Space Administration, and the U.S. Army Corps of Engineers.

**State government** — Most states employ remote sensing specialists to develop land use and natural resource inventories. Such functions are usually housed in the state natural resources or state planning agency, but may also be located in agriculture or transportation agencies.

**Industry** — Remote sensing specialists may be employed in a number of private sector areas, such as large paper, timber, and mining companies, research and development companies working on new technologies and applications, and specialized consulting firms.

**Universities** — Several major universities have associated research institutes that receive government and industry financing for remote sensing research projects. Remote sensing specialists and students are employed to carry out these projects.

### For More Information

American Society for Photogrammetry and Remote Sensing
5410 Grosvenor Lane
Suite 210
Bethesda, MD 20814-2160
(301) 493-0290

## CONSERVATION OFFICER

### Description of the Work

Conservation officers serve to enforce the wildlife, fisheries, and natural resources management laws and regulations of the federal government and state governments. Because much of their work involves dealing with citizens on matters of law violations, conservation officers must have a strong background in law enforcement, as well as in natural resources subjects.

Conservation officers carry out a wide variety of tasks in the performance of their duties, which include: patrolling public lands used for hunting for signs of poaching activity; checking fishermen for proper licenses and inspecting their catch to assure that the numbers are below maximum limits; gathering legal evidence of illegal dredge and fill activity in a protected wetland; responding to citizen reports of suspicious activity involving the trapping of wild game; monitoring activity at a public boat launching ramp and inspecting boats for proper licenses and identification; citing snowmobile drivers for reckless driving or harassment of wildlife; cooperating with local, state, and federal law enforcement agencies in conducting "sting" operations to capture poachers; and visiting public schools to discuss with students the importance of complying with conservation laws and regulations.

Conservation officers may work alone or in teams. This field has traditionally been dominated by males, but in recent years a growing number of women have successfully entered the profession. Individuals considering a career in law enforcement should be comfortable with, and have some familiarity with, handling guns, traps, fishing gear, boats, off-road vehicles, and other equipment and vehicles. In addition, strong diplomacy skills in dealing with people and an ability to work independently are characteristics of a good conservation officer.

## Educational Preparation

Because the competition for conservation officer positions is usually fierce, a bachelor's degree in law enforcement, natural resources, or a related field is generally required in order to have a reasonable chance of attaining a position. However, individuals with substantial law enforcement and/or resource management experience, with some vocational or related training, may also effectively compete for positions.

To prepare for a career as a conservation officer, the following college level courses may prove useful: political science, sociology, psychology, law enforcement, wildlife biology, fisheries, biology, natural resources policy and law, and economics.

## Potential Employers

**Federal government** — Federal agencies directly involved in resource management activities, such as the Fish and Wildlife Service, the National Park Service, and the U.S. Forest Service, regularly employ conservation officers.

**State government** — State natural resources or conservation agencies also employ conservation officers, who are often housed in a special law enforcement division or section.

## For More Information

Contact federal and state natural resources management agencies directly to determine job descriptions and employment requirements.

# ECOLOGIST

## Description of the Work

Ecologists study the relationships between living organisms and the inter-relationships between living organisms and their environment. They conduct research on the habitat requirements, population trends, and associative characteristics of various species of flora and fauna. These scientists provide technical support and advice to managers of natural resource planning and management programs and to commercial interests attempting to locate sites for new development.

In collecting data for specific projects, ecologists often conduct or coordinate field sampling activities. They are involved in the identification and analysis of flora and fauna for various types of habitats. It is common for ecologists to specialize in the study of a specific type of ecosystem, such as forests, wetlands, and open ranges.

Forest ecologists study forestland ecosystems to determine the fragile inter-relationships between trees and other plants, wildlife and fish, and soils, water, climate, and other characteristics of the natural environment. If the ecological characteristics of forests are understood, such lands can be used for multiple purposes (e.g., timber harvesting, wildlife management, watershed protection, and recreation) with minimal disruption of natural systems. In addition, forest ecologists work to reclaim forests damaged by pollution (such as acid rain or acid runoff from a mine) or by prior land uses.

Wetlands ecologists map and classify wetlands, conduct field surveys to verify wetland indicator species, determine the impact of proposed development on wetlands, evaluate the potential of wetlands to be used to treat wastewater discharges from cities and industries, administer wetlands protection regulations for state and local governments, analyze the wetland habitat requirements of endangered species for protection purposes, and plan the development of man-made wetlands to offset the loss of natural wetlands due to land development activities (a process termed "wetland mitigation").

Range ecologists focus on the protection and management of open range lands. Through the study of grassland environments, ecologists determine the extent of use for grazing, recreational, or other activities that can be allowed before degradation of the natural ecosystem occurs. The role of fire on the open range in maintaining new growth of grassy species is a key issue for some range ecologists working in arid areas. They are also concerned with maintaining and protecting the natural populations of wildlife and fish species on the range.

## Educational Preparation

A bachelor's degree in ecological science or a related degree is the minimum requirement for an entry-level ecologist. However, a minimum of a master's degree is preferred by many employers, who may have to use ecological scientists as expert witnesses in court cases or legislative hearings. Consulting companies often look for ecologists who hold doctoral degrees.

High school students interested in the ecological sciences should begin their training early by taking classes in biology, chemistry, and mathematics. College level courses should include: field ecology, botany, wildlife biology, forestry, hydrology, soils, microbiology, advanced chemistry, and advanced ecology.

## Potential Employers

**Federal government** — Federal agencies involved in large-scale land management or regulation employ ecologists. Examples include the National Park Service, the U.S. Forest Service, the Bureau of Land Management, the U.S. Army Corps of Engineers, and the Environmental Protection Agency.

**State government** — State natural resources or conservation agencies often employ ecologists to assist with forest management activities, environmental impact assessment, wetlands delineation and regulation, and related functions.

**Industry** — Large resource-based industries, such as mining and timber companies, occasionally employ ecologists to assist with minimizing the industry's adverse impacts on the environment.

**Consulting firms** — Many ecological scientists are employed by environmental consulting firms to perform wetlands delineations, environmental impact assessments, and siting studies.

**Universities** — Many ecologists holding advanced degrees choose to stay in the academic arena and find work with colleges and universities. They are primarily involved in teaching and research activities.

### For More Information

Ecological Society of America
Center for Environmental Studies
Arizona State University
Tempe, AZ 85187
(602) 965-3000

## WATER RESOURCES SPECIALIST

### Description of the Work

Water resources development specialists, or hydrologists, are concerned with the quantity, flow, distribution, and circulation of water on land, in the soil, and in the atmosphere. The two primary areas of focus for these professionals are the development of a reliable public supply of water and the protection and safety of the public from floods and high water levels.

These scientists collect and analyze large volumes of data on water resources, relying heavily on mathematical calculations. Facts on water resources are compiled for use by engineers, urban and regional planners, government legislative bodies, industry, and other scientists. Hydrologists may even be requested to reconstruct water data from historical records, photographs, and visual evidence such as a high water mark from a record flood. Computer-simulation models are commonly used to predict streamflow levels resulting from heavy rainfall, snowmelt, or other hydrological events.

In addition to studying the water resource itself, hydrologists conduct research on the water supply requirements of communities, industry, agriculture, recreation areas, fish and wildlife habitats, hydropower producers, and other users. From this data, an evaluation of existing and potential future water supplies can be performed.

Hydrologists may become involved in a variety of tasks, including: calculating the rate of streamflow into, and evaporation rates from, water supply reservoirs; projecting the probability of 10-year, 25-year, 50-year, and 100-year flood occurrences on a certain waterway; estimating the level of demand for irrigation, industrial, and municipal water supplies; determining the rate of sediment loading in a reservoir; assessing the current rate of ground water drawdown from a regional aquifer formation; forecasting the impacts of snowmelt and rainfall patterns on a mountain trout stream; predicting the potential for flash floods given a certain set of hydrologic conditions; and evaluating the adequacy of a particular river's flow to drive the turbines at a hydroelectric power facility during low water periods.

Some hydrologists may get involved in studying the effects of new structures within a watershed, such as bridges, dams, and sewer lines. Inspections of the structural integrity of dams and determination of optimal lake levels for particular uses also fall under the hydrologist's scope of activity.

## Educational Preparation

High school students interested in water resources can best prepare for college by taking courses in geography, biology, chemistry, mathematics, and computer operation. College level courses should include hydrology, mathematics, physics, water resources engineering, soil science, geology, hydrogeology, water resources law and policy, economics, and political science.

## Potential Employers

**Federal government** — Federal agencies employing water resources specialists include the U.S. Geological Survey, the U.S. Army Corps of Engineers, the National Oceanic and Atmospheric

Administration, the Bureau of Land Management, the Bureau of Reclamation, the U.S. Forest Service, the Department of Health and Human Services, and the Environmental Protection Agency.

**State government** — State natural resources or conservation agencies employ water resources specialists to assist with various water related programs, such as floodplain management, lake level management, dam inspections, and water supply planning and regulation.

**Local/regional/metropolitan government** — Local, regional, and metropolitan government agencies employ water resources specialists to assist with planning, managing, and regulating the use of public water supplies.

### For More Information

American Water Resources Association
5410 Grosvenor Lane
Suite 220
Bethesda, MD 20814
(301) 493-8600

## SOIL SCIENTIST

### Description of the Work

Soil scientists study, survey, and offer advice on soils for purposes of crop production, land development, site engineering, and conservation. Those who work on soil surveys conduct field tests and classify land on the basis of soil type. Mapped soil survey information is very useful to farmers, engineers, recreation planners, urban planners, watershed managers, and foresters. Soil scientists who work in crop production are concerned with soil fertility, moisture, and cultivation practices. Engineers rely on soil scientists to characterize soils based on their physical properties to support various types of structures or land uses. For conservation purposes, these soil experts work to prevent erosion, protect water quality, and determine ways to use soil as a cleansing filter for sewage or other waste materials.

Like most scientific fields, soil science has several branches or specialty areas, including soil chemistry, soil physics, soil fertility and plant nutrition, soil structure and mechanics, mineralogy, soil genesis, soil biochemistry, and soil microbiology. In recent years, interest in soil microbiology has greatly increased as the biological treatment of contaminated soils has proven to be an effective way to help clean up hazardous waste sites.

Soil conservationists, on the other hand, are concerned primarily with the implementation of conservation practices through planning, design, teaching, and persuasion. Their focus is on assisting landowners, farmers, ranchers, and land developers in preventing the loss of productive soil by wind and water erosion through the implementation of sound soil conservation practices. Planting trees as windbreaks, contour plowing, terracing, crop rotation, reforestation, and the planting of ground cover on exposed soil are all techniques that might be recommended to landowners by soil conservationists. They also provide assistance on water control structures, sediment detention basins, and agricultural waste management practices.

The daily work of soil conservationists may include the following activities: evaluating a parcel of property to make soil and water conservation recommendations; instructing farmers on the benefits and techniques of conservation tillage; reviewing plans of a land developer to ensure that proper erosion control and storm water management measures have been included; assisting Third World countries with evaluating land use and soil characteristics and initiating conservation programs; estimating the deposition of sediment in a protected wetland from a nearby construction site; predicting the long-range effects of weather on soil moisture; designing and overseeing construction of soil and water conservation structures; giving a presentation to a local civic organization on the importance of soil conservation; and preparing a site-specific conservation plan for a farm based on unique topographic, hydrologic, and vegetative characteristics.

## Educational Preparation

A minimum of a bachelor of science degree is required for most soil science and soil conservationist positions. Because few colleges offer degrees in soil conservation, most conservationists earn degrees in soil science, agronomy, forestry, agriculture, or earth sciences.

College level coursework in some of the following subjects is usually helpful to soil scientists and soil conservationists, and many of these classes are required: biology, chemistry, physics, ecology, crop science, agricultural engineering, computer science, soil science, hydrology, forestry, geology, and surveying or drafting.

## Potential Employers

**Federal government** — The federal agencies that most often employ soil scientists are located in the U.S. Department of Agriculture, particularly the Soil Conservation Service and the U.S. Forest Service.

**State government** — State departments of agriculture and natural resources employ soil scientists to assist with technical assistance to farmers and landowners, as well as to support activities on state managed lands.

**Cooperative extension service (university affiliated)** — Land grant universities in each state have government subsidized co-operative extension service programs to assist farmers and landowners with soil conservation and other matters. These programs regularly employ soil scientists as cooperative extension agents.

**Local government** — Municipal and county governments occasionally employ soil scientists to administer their local erosion control ordinances. These local regulatory programs usually require developers and construction companies to implement erosion control techniques on construction sites.

## For More Information

Soil and Water Conservation Society
7515 Northeast Ankeny Road
Ankeny, IA 50021-9764
(515) 289-2331

American Society of Agronomy
677 South Segoe Road
Madison, WI 53711
(608) 273-8080

## REFERENCES

*Soil Scientists*, Brief 276, Chronicle Guidance Publications, Inc., Moravia, NY 13118, December 1986.
*Soil Conservationist*, Brief 201, Chronicle Guidance Publications, Inc., Moravia, NY 13118, March 1990.

# ENTOMOLOGIST

## Description of the Work

Entomologists are specialized biological scientists concerned with the study of insects and their relationship to plants and animals, including man. They develop programs to control insect pests that may endanger trees, food crops, or even human health. The many attributes of insects that are beneficial to man are also studied by the entomologist. Entomologists are introduced here as an example of a specialty area (one of many) within the biological sciences.

Examples of the types of activities entomologists may get involved in include: researching the ecological requirements of a pest species in order to identify environmentally safe means to control their spread; evaluating the effectiveness of various pesticides in controlling explosive mosquito populations; developing integrated pest management programs which incorporate biological controls (e.g., beneficial species feeding on pest species) instead of harmful pesticides; improving fruit and vegetable production by more effectively managing the activity of pollen carrying insects; and educating gardeners on effective techniques for reducing nuisance insect species in a residential area. Entomologists also work to reduce world hunger by developing ways to control losses of crops and livestock caused by insects.

## Educational Preparation

A bachelor's degree in entomology or a related science is the minimum requirement for an entry-level entomologist. However, a minimum of a master's degree is preferred by many employers, who may have to use entomologists as expert witnesses in court cases or legislative hearings. Both chemical companies that manufacture pesticides and specialized consulting firms often look for ecologists who hold master's or doctoral degrees.

High school students interested in entomology need to take biology, chemistry, and mathematics courses prior to college. College level courses for these scientists will include advanced biology, organic and inorganic chemistry, microbiology, ecology, entomology, agronomy, and genetics, along with classes in natural resources management, policy, and economics.

## Potential Employers

**Federal government** — The employers of entomologists in the federal government are primarily found in the Department of Agriculture, particularly the U.S. Forest Service and the Soil Conservation Service.

**State government** — Entomologists are regularly employed by state departments of agriculture.

**Chemical industry** — Manufacturers of pesticides employ entomologists to study the effectiveness of various formulations, as well as to determine the environmental impact of such chemicals. A related industry is the pest control area, which is also a major employer of entomologists.

**Research institutes** — University and industry affiliated research institutes employ entomologists to conduct experimental research and to develop new pest control products and techniques.

## For More Information

Entomological Society of America
9301 Annapolis Road
Suite 300
Lanham, MD 20706
(301) 731-4535

## REFERENCE

Warren, A., *Entomologists*, Brief 243, G.O.E. 02.02.01; 4th ed. D.O.T. 041., Chronicle Guidance Publications, Inc., Moravia, NY 13118, 1989.

# BOTANIST

## Description of the Work

Botanists are concerned with the science of plant biology, a broad field focusing on the entire range of plants and their habitats, from the smallest microscopic pond species to the largest trees in the forest. Botany, like most of the biological sciences, includes a wide diversity of specialty areas, such as plant pathology, breeding, genetics, anatomy, ecology, and microbiology (examples are phycology — the study of algae, and mycology — the study of fungi). In addition, certain areas of applied botany have become separate disciplines of study, such as agronomy (crop science), horticulture (the science of ornamental, fruit and vegetable plants), and forestry (discussed earlier in this chapter).

A botanist might get involved in activities such as: determining the effects of pollution on various types of plants; conducting genetic research on new strains of disease-resistant crops; identifying rare and endangered plants for protection; working to classify and understand newly discovered plants for their usefulness in drugs and medicines; creating, through biotechnology, more nutritious food crops for human and animal consumption; or developing new economic uses for plants as raw material in the production of solvents, paper, resins, rubber, textiles, or construction materials.

Professional botanists may follow a number of career paths. Many are employed as researchers by universities, government agencies, and various industries, where they conduct experiments to further biological knowledge or find useful applications of plant materials to solve problems. A large number of botanists are involved in teaching at the college level, where the work includes both lectures and field activities. Plant-related businesses (e.g., pharmaceutical, seed, and biotechnology companies) are a relatively new, but growing, area of employment for botanical scientists, where their knowledge can be directly applied in the areas of marketing and sales.

A botany degree can also provide a strong background for several niche careers, in which the graduate can meet the unique or specialized needs of a particular employer. One example is the botanist who is employed as a curator or administrator of a botanical garden, natural history museum, or nature center. Others may work as science writers, covering stories about new developments in the field. Computers are playing an increasingly important role in nearly every aspect of science and technology, and botany is no exception. Some specialize in computer programming, where they develop software to aid in research or educational activities. Those with artistic talent may become technical illustrators, using their scientific knowledge to accurately depict botanical subjects. Earning a law degree, after a bachelor's degree in botany, can provide an excellent combination for those interested in environmental, agricultural, or biotechnological legal issues.

Generally, individuals do not choose botany as a career with the expectation of becoming a millionaire. Botanists make adequate to very good salaries, but few get rich. Interesting work, pleasant work environments, and travel opportunities are some of the rewarding aspects of being a professional botanist.

## Educational Preparation

High school students preparing for a career in botany should take courses in biology, chemistry, mathematics, foreign languages, and English. Aside from a core curriculum of college classes in the botanical sciences, undergraduate students should also take courses in the social sciences, general biology, chemistry, physics, computer science, and technical writing.

A bachelor of science degree in botany is the minimum requirement for most professional positions. This degree alone may provide for opportunities as a laboratory technician, educational assistant, or technical assistant in industry or government. As in other scientific fields, additional education and experience leads to a broader range of opportunities. Some positions, such as research positions in industry, may require the minimum of a master's degree or a doctorate degree. To teach at the college level, a Ph.D. is usually required. The combination of a bachelor's degree in botany with an advanced degree in another field, such as journalism, law, art, or business administration, can lead to specialized careers in these areas (see Chapter 6).

## Potential Employers

The primary employers of botanical scientists are educational institutions, government agencies (federal and state), and plant-related industries. Educational institutions include both community colleges where general botany courses are taught, and four-year colleges and universities where specialized teaching and research positions exist. Government positions at the federal level are mostly in the Department of Agriculture (U.S. Forest Service, National Arboretum, Soil Conservation Service, and others) and the Department of the Interior (National Park Service, U.S. Geological Survey, Bureau of Land Management, and others); while botanists are also hired by the Public Health Service, State Department, NASA, the Smithsonian Institution, and the Environmental Protection Agency. The types of industries that hire plant biologists include: pharmaceutical, petrochemical, chemical, lumber and paper, seed and nursery, fruit growers, food processing companies, fermentation industries (such as breweries), biological supply houses, and biotechnology firms.

### For More Information

The Botanical Society of America
Department of Botany
Ohio State University
1735 Neil Avenue
Columbus, OH 43210

## REFERENCE

*Careers In Botany*, The Botanical Society of America, c/o Department of Botany, Ohio State University, Columbus, Ohio 43210, undated.

## NATURAL RESOURCE POLICY ANALYST AND PROGRAM ADMINISTRATOR

### Description of the Work

Natural resource policy analysts perform a wide variety of functions related to the development or modification of government policies, laws, and regulations. These environmental professionals use political and legal skills, along with a technical understanding of their area of specialty (e.g., forestry, water resources, wildlife, etc.) to establish or influence public policy.

The duties of the policy analyst might include: evaluating proposed legislation; estimating budgets for natural resources management programs; determining the impact of certain regulations on various affected parties; conducting policy research; building coalitions to support proposed laws; conducting economic impact analyses of government subsidy programs; planning strategies for policy development related to controversial natural resource issues; drafting or revising proposed policies, legislation, or regulations; or preparing position papers for presentation at public hearings or committee meetings. In addition, they get involved in estimating required staffing levels for proposed government programs; determining popular opinion on specific issues through survey data; soliciting input on proposed policies; facilitating consensus of opinion in committee situations; consulting individually with various interested parties; developing and evaluating alternative policy scenarios; predicting political positions by various parties on proposed government or legislative actions; analyzing regional differences in program impact; identifying critical geographic area positions on issues; and developing and maintaining liaisons with federal, state, and local agencies.

Natural resource economists, a special type of policy analyst, focus on the economic aspects of natural resource management programs. The economist's duties might involve: conducting cost/benefit analysis on proposed government or industry programs; developing sophisticated economic impact assessment procedures for certain natural resource management actions; projecting economic growth of resource based industries; determining the optimal plant and harvest rotations for the timber industry; or estimating the economic benefits to a coastal community from sport or commercial fishing activities.

Program administrators are responsible for the day-to-day operations of government natural resources management programs (e.g., coastal zone management, erosion control, forest management, etc.). The administrative duties often include: hiring, firing, supervising, and training staff; solving personnel and other problems; developing proposed annual budgets; proposing policy; coordinating activities with related government programs; interpreting laws and regulations;

conducting public relations and education activities; scheduling and assigning work; monitoring progressiveness and effectiveness of programs; and identifying and implementing corrective actions.

## Educational Preparation

High school students interested in policy analysis and public administration in the natural resources field should take courses in biology, chemistry, mathematics, social studies, civics, English, and computer operation. College level courses should include at least some of the following: political science, natural resources policy and law, economics, ecology, urban and regional planning, sociology, psychology, survey research methods, public administration, and a series of classes in a natural resources specialty area (e.g., recreation, fish, law enforcement, land use, etc.).

## Potential Employers

**Federal government** — All federal resource management agencies, primarily those in the Departments of Agriculture and Interior, employ policy analysts and program administrators.

**State government** — State natural resources and agricultural agencies also employ policy analysts and program administrators.

**Industry** — Large resource-based industries (e.g., mining, pulp and paper, chemical, utility, petroleum, etc.) employ policy analysts to track legislation and monitor government regulatory programs which affect their operations.

**Research institutions** — Several "think tank" organizations affiliated with universities or industries employ policy researchers to study, analyze, and predict legislative and regulatory needs and trends.

**Environmental interest groups** — These organizations employ policy analysts to oversee and monitor the operations and activities of government agencies to assure their compliance with their own policies and regulations.

## For More Information

Contact employers directly for job descriptions and employment requirements.

# OCEANOGRAPHER

## Description of the Work

The interdisciplinary study of the characteristics and processes of oceans and seas is the domain of the oceanographer. Such study is important for determining the environmental impact of dumping of municipal and industrial wastes, the identification and protection of saltwater habitats and species, the discovery of economically viable mineral deposits or oil fields on the ocean floor, and the dynamics of waves, currents, and tides. Specialties within the field of oceanography include: physical oceanography, which deals with waves, currents, tides, weather relationships, temperature differences, ice formation and flows, and related physical phenomena; geological oceanography, which concerns itself with the contours of the ocean floor and the geologic composition and movement of the submerged ridges and mountains; chemical oceanography, which focuses on the constituents (natural elements, as well as contaminants) and interactions within the water and between the water, marine life, and the atmosphere; and perhaps the most popular discipline, biological oceanography (commonly known as marine biology), which involves the study of marine plants and animals and their interactions.

Oceanographers usually begin their careers as research assistants or technicians on research vessels during the summers of their college years. They generally collect data and take samples with various types of instruments and recording devices, and analyze the data in a laboratory or research center. Professional oceanographers also spend some time at sea on research vessels. Occasionally, they may dive underwater, either in submersible vessels or in scuba gear, to collect data or samples. Sampling instruments are used to take water quality, marine life, or bottom sediment samples. Other instruments measure current flow velocities, temperature, gravity, depth, and other characteristics of the ocean. Underwater cameras are commonly used to make visual observations of marine life or the ocean floor. After each research voyage, substantial amounts of data are brought back to shore for analyses and interpretation.

Working in land-based research centers, oceanographers plan research projects, conduct library research, interpret data, record the details of sampling and testing procedures, develop theories from analyzed data, write scientific reports, and make presentations to scientific societies. Oceanographers also get involved with laboratory work where activities include: conducting chemical tests on water samples; measuring, photographing, and cataloging marine flora and fauna; classifying sediments and minerals; preparing maps and charts to plot sampling locations; and using computers to sort and manipulate raw data and to run simulation models. Senior oceanographers may have additional administrative responsibilities such as supervising a staff and managing projects and research facilities.

Historically, most oceanographers were employed by government and academic institutions. However, because of increased commercial interest in economic development of ocean resources, private industry is hiring increasing numbers of oceanographers. However, aspiring oceanographers should expect intense competition for the higher paying and most interesting positions, whether with the government, industry, or in academia.

## Educational Preparation

High school students interested in oceanography should take classes in biology, chemistry, geography, mathematics, and computer operation. College level courses should include geography, marine biology, microbiology, chemistry, geology, physics, and mathematics. Additional coursework in meteorology, hydrology, coastal geology, engineering (environmental, civil, structural, etc.), and other specialized areas of interest will also prove useful.

The minimum educational requirement for professional oceanography positions is a bachelor's degree in science. Graduate degrees are required for those interested in teaching and research positions.

## Potential Employers

**Federal government** — The following federal agencies employ oceanographers: the Department of Defense (Navy, Army Corps of Engineers), the National Marine Fisheries Service, the U.S. Geological Survey, the U.S. Fish and Wildlife Service, the National Oceanic and Atmospheric Administration, the Department of Energy, and the U.S. Coast Guard.

**State government** — Oceanographers are also employed by state departments of natural resources (in coastal states).

**Industry** — The industries which employ oceanographers include those involved with ocean mining, geophysical exploration, petroleum, shipping, and commercial saltwater fishing.

**Academia** — Oceanographers are also employed by universities with strong oceanographic research and educational programs, such as Scripps Institution of Oceanography (University of California at San Diego), Woods Hole Oceanographic Institution (Woods Hole, Massachusetts), Lamint-Doherty Geological Observatory (Columbia University), and Rosenstiel School of Marine and Atmospheric Science (University of Miami).

**Nonprofit organizations** — Some oceanographers are employed with the larger environmental interest groups that sponsor research activities in support of policy positions on matters related to the oceans.

### For More Information

Oceanic Society
218 D Street S. E.
Washington, D.C. 20003
(202) 544-2600

American Oceanic Organization
c/o Department of Commerce
1825 Connecticut Avenue N.W.
Washington, D.C. 20235
(202) 673-5140

## REFERENCES

"Oceanographers", Chronicle Guidance Brief 200 (4th ed. D.O.T. 024.; G.O.E. 02.01.01), Chronicle Guidance Publications, Inc., Moravia, NY 13118, November 1987.

"Oceanography", by Karen Hartley, article appearing in *Future Careers* magazine, vol. 2, no. 2, Fall/Winter 1990, p. 40-41, published by Career Information Services, Inc., 70 Seaview Avenue, Stamford, CT 06902, (203) 353-7066.

## GEOGRAPHER

### Description of the Work

The field of geography is concerned with the characteristics and relationships of space, location, and place; in the broader context of people and culture; and the interactions between the physical and human environment. Geographers might specialize in physical geography (focusing on landforms, soils, hydrology, climates, and vegetation), economic geography (concerned with the locational aspects of economic activity, trade, production, markets, and employment), social or urban geography (studies populations, where people live, why they live there, the distributions, densities, and characteristics of people in an area), political geography (dealing with governmental boundaries, voting patterns, defense considerations, and conflicts), historical geography (analyzing geographical aspects of the past), or environmental geography (concentrating on the relations between humans and the ecosystem). Geographers may work on the global, national, regional, state, or local scale in analyzing the spatial aspects of human activity and natural phenomena.

In their study of the spatial relationships of the physical and human environment, geographers get involved in widely differing types of activities. For example, studying the dynamics of world population growth and food shortages, documenting migration patterns of people within the U.S., inventorying natural resources within a state and analyzing changes over time, determining the best location for a factory or corporate headquarters, or delineating the distribution of housing projects within a metropolitan area, are all tasks taken on by the geographer.

The techniques of geographic study involve the collection, analysis, and interpretation of large volumes of data, traditionally gathered through field observations. Today, these field observations are greatly enhanced by the use of aerial photography and satellite imagery. Computers are used to perform statistical analyses of geographic data. Often the results of data analyses are plotted on maps or other geographical representations. The plotting of these large volumes of data is increasingly done through the use of computer-based geographic information systems.

Professional geographers find employment in teaching (at all levels), research, map and journal publishing, and in numerous positions in government and industry, often with titles that do not include the word *geographer*. They work in geographic subdisciplines such as cartography, remote sensing, and urban and regional planning (each of these is addressed elsewhere in this book), as well as in local housing/community/economic development, industrial and retail site location, resource conservation and management, land use planning, climatology, international trade and economics, transportation analysis, and in other specialty areas. Federal, state, and local government agencies employ large numbers of geographers. The employment of geographers by businesses has increased dramatically in recent years, as geographic approaches to solving business problems is becoming more popular. Academia also employs many geographers for teaching and research positions, but students should expect the competition to be strong.

## Educational Preparation

High school students interested in geography should take classes in social studies, civics, biology, mathematics, drafting, and political science. At the college level, students interested in the environmental aspects of geography should consider taking courses in physical geography, regional geography, biogeography, cartography, remote sensing, economics, ecology, urban and regional planning, hydrology, soils, forestry, botany, water resources management, and natural resources and environmental law.

The minimum educational requirement for professional geographers is a bachelor's degree in geography. For the best opportunities at the higher paying management positions, students should consider completing graduate study at the master's level. A master's degree is considered the minimum for teaching at the college level, and for most research positions. However, many colleges and universities will require a Ph.D. for teaching and senior research positions.

## Potential Employers

**Federal government** — The federal government employs geographers in a wide variety of agencies, although the position titles may be related to program or administrative functions rather than *geographer.* These agencies include the Environmental Protection Agency, the National Park Service, the U.S. Forest Service, the Bureau of Land Management, the National Oceanic and Atmospheric Administration, the Department of Housing and Urban Development, the U.S. Army Corps of Engineers, the Department of Health and Human Services, and the Department of Transportation. In addition, federally authorized agencies, such as the Tennessee Valley Authority, employ geographers.

**State government** — States also hire geographers in agencies involved with environmental protection, public health, housing, transportation, commerce, and agriculture.

**Sub-state regional and metropolitan planning agencies** — These multi-city and multi-county agencies employ geographers to analyze spatial patterns and relationships to help plan for growth and development in specific geographic areas.

**Local government** — Individual cities, townships, and counties employ geographers to manage and carry out local planning, zoning, and community development programs.

**Consulting firms —** Planning, landscape architecture, architecture, and engineering consulting firms employ geographers to assist on large research projects, as well as to offer technical siting and location assistance to communities and businesses.

**Universities and schools** — Many geographers are employed as professors and researchers at college universities, or as teachers in public and private school systems.

**Other potential employers** — Geographers are also employed by land developers, banks, public utility companies, map publishing companies, and marketing firms.

### For More Information

Association of American Geographers
1710 Sixteenth Street N.W.
Washington, D.C. 20009-3198
(202) 234-1450

American Geographic Society
156 Fifth Avenue
Room 600
New York, NY 10010
(212) 242-0214

## REFERENCES

*Why Geography? Answers from the Association of American Geographers*, a brochure prepared by the Association of American Geographers, 1710 Sixteenth Street N.W., Washington, D.C. 20009, (202) 234-1450.

Huke, R. E. and Malstrom, V., *Geography As A Discipline*, a brochure published by Association of American Geographers, 1710 Sixteenth Street N.W., Washington, D.C. 20009, (202) 234-1450.

*Geographers*, Chronicle Guidance Brief 155 (4th ed. D.O.T. 029.; G.O.E. 02.01.01), Chronicle Guidance Publications, Inc., Moravia, NY 13118, December 1987.

CHAPTER 8

# Non-Degree Technical Environmental Careers

# Non-Degree Technical Environmental Careers

Not all important environmental work is performed by professionals holding bachelor's or graduate degrees. There are tremendous opportunities for trained technicians to play key roles in environmental protection and natural resources projects and programs. The majority of non-degree technical positions (by "non-degree" I mean less than a four-year bachelor's degree) involve specific types of data collection and analytical tasks, or on-site field work, under the supervision of a degreed environmental professional. Nearly every area of environmental work employs some type of technician.

## NATURAL RESOURCES MANAGEMENT TECHNICIANS

This group of technical specialists includes those who work in wildlife management, fisheries management, parks and recreation, and land use management and control. In the areas of fish and wildlife management, technicians perform a wide variety of hands-on field work, such as: handling game animals for research or restocking purposes; collecting specimens to study diseases; performing maintenance work on special facilities and equipment at game farms and fish hatcheries; transporting and releasing animals; conducting field surveys of species in the wild; evaluating the condition of habitats; and capturing and removing problem animals or species when they have become pests. Park rangers carry out assignments ranging from law enforcement to park maintenance to public relations. Land use management types of programs often hire planning technicians and cartographers. Planning technicians assist with interpreting data from aerial photographs, performing "groundtruth" inspections of geographic data, compiling statistics, and related tasks.

The majority of technician employment opportunities in natural resources management are with federal, state, and local government agencies and occasionally with academic institutions involved in field research activities. The qualifications for these technicians vary with the employer. Potential employers should be contacted directly to determine specific requirements for employment.

## ENVIRONMENTAL TECHNICIANS

Non-degree technical careers in the environmental protection arena cover a broad range of work situations, from laboratories and industrial plants to offices and waste treatment facilities. The following discussion identifies several types of these technical positions, although it is certainly not exhaustive or inclusive of the subject.

### WASTEWATER TREATMENT PLANT OPERATOR

Wastewater treatment plant operators are responsible for the continuous running and maintenance of both publicly owned treatment works (POTWs), which handle municipal sewage, and industrial wastewater treatment and pretreatment facilities. State governments require wastewater treatment plant operators to be trained and certified before being licensed to run these types of systems. This requirement is a manifestation of the high level of public responsibility placed upon the wastewater treatment plant operator.

The training that wastewater treatment plant operators receive relates to the physical (e.g., solids removal, aeration, oil/water separation, etc.) and chemical (e.g., pH adjustment, coagulation, flocculation, chlorination, etc.) aspects of treatment systems. There are different classes of operator licenses that are differentiated by the sophistication of the treatment technologies. The higher the license class, the stricter the training requirement.

In addition to training on the operation of a wastewater system, operators also receive training on the regulatory aspects of wastewater treatment. Every wastewater treatment facility operates under a state or federal discharge permit (called an NPDES permit, which stands for National Pollution Discharge Elimination System, from the original clean water program) which allows a discharge into a stream, or under a local permit allowing hookup to a POTW facility. There is also an industrial wastewater *pretreatment* program where industrial plants are allowed to hookup to a POTW system only after the industrial wastewater has received certain types of pretreatment. Within these permits are effluent limitations which set forth the maximum level of various pollutants that are allowed to be discharged. Wastewater treatment plant operators need to understand these effluent limitations because they directly relate to the performance of a treatment system. If a wastewater treatment system violates its NPDES permit, the facility is subject to fines and other enforcement action.

A career as a wastewater treatment plant operator offers "grassroots" and "hands-on" experiences where the results of personal effort are seen directly, and on a daily basis. Municipalities and industry are the employers of licensed wastewater treatment plant operators.

## Incinerator Operator

The incineration of both solid and hazardous waste is becoming increasingly accepted by scientists, engineers, and regulatory agencies as an environmentally sound means of disposing of these materials. Municipalities struggling with limited landfill capacities to dispose of solid waste are building waste-to-energy facilities which not only burn the solid wastes, but also recapture the heat to produce energy. As more and more of these incineration facilities are constructed, there is an increasing need for qualified operators.

The job of the incinerator operator is to efficiently and effectively run the system within regulatory limitations and sound management practices. The operation of incinerators must meet the requirements of an air quality permit, a water quality permit (if there is a wastewater stream from the process), solid waste or hazardous waste permit or manifest for ash disposal, and possibly other limitations relating to noise, dust control, underground storage tanks, emergency response, and/or other items. The operation of waste incinerators is under constant scrutiny by the public and regulators, making the already heavy responsibilities of the operator even more important.

## Landfill Operator

The traditional way to dispose of municipal solid waste and nonhazardous industrial waste has been to deposit it in a sanitary landfill (a managed site) or a dump (an uncontrolled site). Although a tremendous amount of effort is being exerted by government and industry to reuse and recycle solid waste, landfills are, and will continue to be, the ultimate depository for nonhazardous waste materials. The operation of landfills today is much more complex than even just a few years ago, presenting a challenge to those involved in managing them.

Landfill operators are responsible for ensuring that all aspects of the landfilling activity are performed in an environmentally sound and legal manner. For example, refusing to accept certain types of industrial wastes that have been hauled to a landfill, because they potentially contaminate the ground water underneath the site, and keeping the landfilled wastes covered with earth on a daily basis to prevent problems with scavengers (such as birds and rats), are some of the duties of the landfill operator. At many landfill sites, certain parts of the operation may have to be closed (either because they are full or because they may have contributed to soil or ground water contamination), while other portions may have to be expanded (new requirements will probably include impermeable liners and ground water monitoring). Landfill operators at the larger facilities are likely to be assisting with these activities.

## Recycling/Reclamation Plant Operator

Public interest and changing regulations are creating a growing demand for recycling and reclamation facilities to handle newspapers, steel and aluminum cans, glass, certain types of plastics, as well as a wide variety of industrial waste. These facilities will need competent operators to keep the processes efficient and economically viable. Opportunities in this area will likely increase substantially through the 1990s.

## Hazardous Material Transporter

The transportation of hazardous chemicals and waste materials requires special handling and safety measures. These substances, whether they be transported by airplane, ship, train, truck, or other means, must be contained and moved in ways which prevent or minimize the possibility of accidents and spills. The people who transport hazardous materials need to know how to protect themselves from exposure, how to safely operate vehicles and vessels, and how to respond to accidents when they occur. In addition, they must know hazardous material labeling and manifesting requirements.

The need for hazardous material transporters will undoubtedly increase as these substances are shipped from manufacturer to user, from user to treatment, storage, or disposal facilities, and to recyclers and reclaimers.

## Hazardous Waste Site Worker

There are numerous opportunities for technicians trained for work on hazardous waste sites. These positions involve operational work at licensed hazardous waste treatment, storage, or disposal facilities, and field work at Superfund, RCRA, or state program contaminated site cleanup projects.

Hazardous waste technicians are thoroughly trained in personal protection techniques as well as in hazardous waste handling methods. This work may require an individual to wear a chemical-resistant suit, gloves, helmet, and in extreme circumstances, a self-contained breathing apparatus or other type of respirator. These technicians use special tools to locate, sample, and remove buried steel drums, underground storage tanks, and other possible sources of contamination. They may also run equipment to excavate contaminated soils from cleanup sites.

The demand for hazardous waste technicians will certainly increase through the decade, as many federal Superfund projects, state-funded cleanups, and industry-sponsored cleanups are moving from the site investigation and evaluation stage into actual remediation activity.

## Remediation System Operator

Another type of technical work associated with the cleanup of contaminated sites is the operation of soil and ground water remediation systems. This type of technician is trained to operate and maintain treatment systems to reduce the content of hazardous constituents in contaminated soil or ground water. Depending on the particular type of technology used, remediation system operators may need to be certified or licensed in the same way as a wastewater treatment plant operator. For example, to clean up contaminated ground water at a site, a system might be installed that pumps the water to the ground surface for treatment in a small on-site system. In this case, the operator would likely be required to be licensed to run the ground water remediation system.

Similar to the hazardous waste technician discussed earlier, the demand for remediation system operators will increase proportionately with the increase in cleanup activity.

## Laboratory Technician

Virtually every environmental protection project or program involves, at some point, the analysis of samples of the air, water, ground water, soil, or waste liquids, sludges, and solids. It is critical for environmental scientists and engineers to know which pollutants they are dealing with, and in what concentrations. This chemical analysis takes place in environmental laboratories, under the supervision of a laboratory scientist. However, in many cases, most of the testing procedures are performed by laboratory technicians. Environmental laboratory technicians calibrate laboratory equipment, prepare samples for testing, follow detailed and often complex procedures for measuring pollutants, and record the results of the tests performed. In some cases, laboratory technicians may go to a facility or project site to take the needed samples.

Many federal and state environmental agencies have their own environmental laboratories. However, the best employment opportunities for laboratory technicians are with one of the growing number of commercial environmental laboratories. These commercial laboratories are responding to the increasing demands of industry for analytical support to their ongoing compliance monitoring for air, water, and wastes, as well as to site remediation projects.

## Emergency Responder

Emergency responders are technicians trained primarily to respond to chemical spills and accidents. They are intimately familiar with personal protection techniques as well as with spill containment and cleanup methods. Many chemical emergency responders were first trained as fire fighters, either within industrial plants or with community fire departments.

Chemical emergency responders are employed by large manufacturing plants, communities (either with the local fire departments or with specialized hazardous materials response teams), and commercial spill response companies who work under contract with government agencies and industries.

### Geology Technician

Geology technicians assist professional environmental geologists with a wide variety of tasks, many of which are performed in the field. Examples include: conducting site surveys using geophysical testing equipment and soil vapor meters; drilling ground water wells for investigative sampling or monitoring; drilling soil cores; taking soil and ground water samples; preparing samples for shipment to a laboratory; filling out chain-of-custody papers for laboratory samples; decontaminating field equipment; recording well locations, depth to ground water, and other information; preparing field reports; and transporting field equipment to project locations.

Any employer of an environmental geologist would also be a potential employer of a geology technician. Federal and state government agencies hire geology technicians to assist with contaminated site investigations. Industries use such technicians to perform routine ground water monitoring activities associated with their production operations. Environmental consultants and engineering companies use geology technicians to perform literally all of the tasks mentioned in this section.

The demand for geology technicians appears to be strong and steadily increasing as more and more contaminated sites are discovered and reported.

## ENGINEERING-RELATED TECHNICIANS

Another category of non-degree technical careers deals with engineering-related functions and activities. These opportunities generally involve measuring, calculating, plotting, drafting, or otherwise collecting and manipulating technical information.

## Engineering Technician

Working under the supervision of a professional environmental engineer, engineering technicians get involved in: drafting detailed engineering plans from conceptual engineering designs; performing calculations to determine pipe sizes, pump sizes, stack heights, and a myriad of other necessary design details; collecting and evaluating information on pollution control and waste treatment equipment supplied by various vendors; and entering engineering information on computer data bases. It is the responsibility of the engineering technician to assist professional engineers in being more productive.

Some of the tasks performed by engineering technicians require special training, while other tasks can be learned on the job. Employment opportunities for engineering technicians are steadily increasing as the national shortage of qualified engineers keeps pressure on working professional engineers to get more done in less time.

## Computer-Assisted Design (CAD) Operator

Computer-assisted design (or CAD) technology has revolutionized the engineering field by providing a means to quickly prepare, modify, and finalize both conceptual and detailed engineering drawings. CAD system operators are specially trained to effectively and efficiently utilize the capabilities of the powerful CAD software to plot and modify complex drawings.

Both engineering companies and government agencies hire CAD operators to develop engineering drawings, site sketches, maps, and other graphic images. The market for proficient CAD operators is and will continue to be strong in the environmental arena, as new waste management facilities are designed and contaminated sites are cleaned up.

## Surveyor

Land surveyors layout and verify property boundaries and elevations, a service often utilized on large-scale environmental projects. They receive specialized training in the use of transits and other surveying equipment, and in plotting boundaries and elevations on maps and engineering drawings. Surveying involves outdoor field work as well as office time to prepare the survey documents.

Land surveyors are hired by a wide variety of employers, including engineering companies, mortgage companies, government agencies, companies owning large acreages of land. In addition, many surveyors work independently on a contract basis, or with a group of surveyors operating as a surveying company.

### CARTOGRAPHY TECHNICIAN

Cartographers create and modify maps for general use, as well as for an entire spectrum of specific purposes, such as for natural resources and environmental protection. Using many types of geographic data, cartographers create base maps as a foundation upon which to plot virtually any type of information (e.g., wildlife population densities, regulated wetlands, the distribution of wastewater effluent outfalls along a river, 100-year floodplains, the location of endangered species, or the lateral extent of known ground water contamination within a county.

The use of drafting tools by cartographers has largely given way to highly sophisticated geographic information systems (GIS). These GIS computer systems allow cartographers to quickly develop new base maps and to load and plot tremendous amounts of geographic data about the area covered by the map. By using GIS, cartographers can overlay various types of information over base maps to analyze specific aspects of a site (e.g., the proximity of a proposed wastewater treatment plant to a sensitive fish spawning area in a river).

Today's cartographers need to receive training in the operation of GIS systems, and to further develop their proficiency and skill through practice. Highly skilled and creative cartographers should find good employment opportunities with companies and government agencies using GIS.

## GETTING STARTED

To learn more about specific technical career opportunities, begin by contacting potential employers to determine their particular requirements for technical positions. Such requirements may range from simply having a high school diploma and good employment credentials, to a very specific associate's degree or technical certification. Once the educational requirements have been identified, contacting training institutions which offer the relevant instructional programs is the next step. Some employers may allow a job candidate to begin work while completing technical training, and some may even provide tuition reimbursements.

## SUMMARY

Not everyone interested in environmental work is cut out to earn a bachelor's degree. However, this discussion has highlighted some interesting technical career options within the environmental field that can be successfully entered without benefit of a four-year college degree. There are certainly other existing technical career opportunities that were not mentioned in this chapter. It may take a certain amount of hard thinking and personal motivation, but satisfying technical careers in the environmental field are available to those who seek them out.

# CHAPTER 9

# Opportunities for International Work

# Opportunities for International Work

Virtually every environmental career identified in this book can be successfully pursued on an international basis. This is particularly true for the scientific and engineering professions, where technical problem solving can be easily transferred across national, political, and cultural lines.

Environmental professionals are, and will continue to be, in strong demand in both industrialized nations and developing Third World countries. With the European Economic Community agreement going into effect in 1992 (where governmental regulations will become more standardized and trade barriers will be modified or removed), and with the opening up of the post-Communistic Eastern European markets, opportunities for industrial and governmental environmental work are expected to substantially increase. As relations between the U.S. and the former Soviet Union (now the Commonwealth of Independent States, CIS) continue to warm, the free flow of information and technical assistance on environmental protection techniques will be high on the priority list for the CIS, a region that is just beginning to address these issues. The countries of the Pacific Rim, many of which currently have international trade surpluses (e.g., Japan, Taiwan), are actively seeking technical expertise to help solve an entire spectrum of environmental problems, many the result of years of neglect. Several of the war-torn countries of the Middle East are in need of systems to provide clean, potable water, effective wastewater treatment, and a sanitary means to dispose of municipal solid wastes. The Third World countries of Latin America and Africa, as they continue to develop, may be able to avoid the pitfalls of industrialization through the implementation of sound environmental planning and management programs. The demand for technical assistance in these countries will grow more slowly than that for more industrialized nations because of a lack of financial resources.

Who employs environmental professionals for international work? Governments, multi-national corporations, international organizations, and large environmental interest groups all have occasional needs for environmental specialists to work in international settings, either working domestically or in foreign countries. The federal government of the U.S. and the governments of foreign nations both hire American citizens to perform environmental work. The U.S. Environmental Protection Agency has offices specifically designated to coordinate international activities. Similar agencies exist in foreign countries which may have need for personnel trained in particular environmental specialties. Large multi-national corporations who manufacture or distribute products overseas have the same need for environmental expertise as domestic companies. Environmental specialists might be employed to minimize pollution at foreign production facilities or to perform product stewardship activities (e.g., providing proper handling, safety, and disposal procedures for chemical products containing hazardous substances) for foreign customers. Another employer category is international organizations, such as the United Nations and the World Bank. Developing proposals for environmental projects, project management, preparing technical assistance publications, or drafting model environmental policies are all examples of work offered by these organizations. The larger environmental interest groups, such as the National Wildlife Federation, the National Audubon Society, and the Nature Conservancy also offer limited international employment opportunities. The work might involve campaigning to ban the use of ozone-destroying chlorofluorocarbons, managing a wildlife refuge, or purchasing a threatened rainforest ecosystem for ultimate protection.

Educational preparation for a career in international environmental work involves the same training requirements as those discussed in the various chapters of this book, with a few additions. First and foremost is the learning of one or more foreign languages. Fluency in a foreign language provides prospective career seekers with a strong credential for conducting business in the international arena. Among the more useful foreign languages being taken by students are Japanese, German, Spanish, and Russian. In addition to a foreign language, educational training in the culture and politics of the particular country or region of interest is also important.

Beyond educational training, there are a number of other activities that the environmental job hunter should consider pursuing to increase the level of personal exposure to a particular country's language and culture. Becoming active with an international studies center at a major university is one approach. Another is to take advantage of every opportunity to travel to the foreign country of interest (e.g., vacations, educational trips, peace corps service, church missionary work, etc.). Making contacts at international organizations that are involved in a particular country may yield opportunities to assist on special projects on either a voluntary or paid basis. Another way to make professional contacts is to identify foreign-owned companies that are operating in the U.S., and express your interest in environmental work in that country.

## For More Information

Smithsonian Institution
1000 Jefferson Drive S.W.
Washington, D.C. 20560
(202) 357-2700

United Nations Environment Programme
United Nations
Room DC2-0803
New York, NY 10017
(212) 963-8138

World Environment Center, Inc.
419 Park Avenue
Suite 1404
New York, NY 10016
(212) 683-4700

The World Bank
1818 H. Street N.W.
Washington, D.C. 20433
(202) 477-1234

"Opportunities Overseas"
International Development Intern Program
Office of Personnel
Agency for International Development
Washington, D.C. 20523

Also, contact the larger multi-national corporations about international employment opportunities in the environmental field. Universities have international centers which have information on government, business, and academic addresses in various countries. Approach international consulting firms about current work overseas.

CHAPTER 10

# Employers of Environmental Professionals

# Employers of Environmental Professionals

There are literally dozens of types of employers for environmental professionals, each with their own set of peculiarities, advantages, and shortcomings. In this chapter, several general categories of employers will be examined to highlight the special characteristics associated with each for the prospective job candidate.

## GOVERNMENT

For numerous individuals employed in the environmental field, government employment best matches their personal beliefs and idealism. As a public servant, you are employed to develop and implement government policies, laws, and regulations, and to manage natural resources for the common good (a basic tenant of environmentalism). These governmental positions carry the responsibility of official authority, which is to be used wisely, equitably, and appropriately.

Government workers are commonly referred to as "bureaucrats" by those who are affected by their actions. This term is often used in a negative tone by industrial representatives, environmental groups, citizen organizations, and even legislators, whose laws are implemented by these public officials. A "thick skin" is required for government work. New employees will be challenged to assist government in regaining its former level of public respect.

### *FEDERAL GOVERNMENT*

Environmental professionals employed by the federal government perform a broad spectrum of tasks, from a "big picture" or "global" perspective. Much of this type of work supports public policy development and implementation, such as proposed legislation to reduce sulfur dioxide emissions or to protect a specific wilderness area. Another important aspect of environmental work with the federal government is basic and applied research upon which laws, regulations, and programs are based.

In many of our country's environmental protection and natural resources management programs, federal government funds are passed through to state and local agencies for implementation (e.g., Clean Water Act permit program and Land and Water Conservation Fund for recreational land acquisition). A large number of administrative positions are dedicated to distributing, coordinating, and monitoring these grants to state agencies and local governments. Another key function of federal agencies is to provide technical assistance to the states. This often takes the form of preparing guidance manuals, conducting training courses, and facilitating the flow of technical information between the agencies.

## The "Pros"

Federal employment offers several advantages desired by a large number of environmental professionals. Some of the more common work-related advantages include: opportunities to influence public policy and programs on a national scale; higher levels of responsibility sooner in a person's career than in other types of employment; and exposure to an extensive network of environmental professionals at the federal, state, and local levels of government, as well as at environmental interest groups and trade associations. In addition, many people find the legislative and political processes an extremely interesting and challenging aspect of their careers.

On the personal level, employment with the U.S. government offers usually good pay, good medical and dental benefits, a certain degree of job stability and protection under the Civil Service system, and regular training and professional development opportunities. Basic full-time federal employment also includes sound retirement benefits and the full complement of paid government holidays. In addition, federal employees have the opportunity to move to various areas of the country through transfer and promotional opportunities, without having to change employers.

## The "Cons"

Although there are many advantages to working for Uncle Sam, the aspiring environmental public servant must also evaluate the commonly perceived "down sides" to determine whether or not they are important enough to warrant looking elsewhere for employment.

Some of the common complaints of federal employees, as heard by the author over a period of several years, include: being too far away from the action of state and local agencies, and industries; being frustrated by the delays and political posturing inherent within a large bureaucracy; the majority of employment locations being limited to only the largest cities (with the exception of agencies managing large land areas, or who have special research facilities); program success often depending on national level politics beyond the individual's control or influence; professional advancement sometimes seeming to be based on seniority or politics rather than on personal performance or merit; and modest pay relative to salaries offered by industry for similar work.

## *State Government*

Professional environmental work at the state level of government has a slightly different focus than at the federal level. Although still involved in the development of public policy and legislation, state officials may become much more actively involved in the enforcement of regulations (both federal and state) and providing technical and funding assistance to city and county agencies.

For many federal environmental protection and resource management programs, the implementation authority is passed on to the states along with federal funding. Examples include the National Pollution Discharge Elimination System permit program and others under the Clean Water Act, the Resource Conservation and Recovery Act programs for solid and hazardous waste management, the permitting and enforcement programs incorporated into a State Implementation Plan under the Clean Air Act, the Land and Water Conservation Fund for purchasing and developing recreational lands, the Pittman-Robertson funds for wildlife management, and the project/coordination/technical assistance funding under the Coastal Zone Management Act. Because of this situation, state environmental officials are on the "front lines" of program management, attempting to carry out federal mandates through program planning and scheduling; collecting and analyzing large volumes of technical data; dealing directly with the public at hearings and through the handling of complaints; conducting field investigations and inspections; and coordinating and communicating on activities with federal and local agencies, business and industry, environmental interest groups, trade associations, citizen organizations, sportsmen's groups, and the news media.

Much effort within state environmental and natural resources programs is expended in providing technical assistance and administering project or program grants to local agencies. In the former instance, city and county officials look to the state for technical guidance and advice on complex projects, such as developing a five-year solid waste management plan for a metropolitan area. The state official is often perceived as "the expert" or "the authority" in this role. As such, it is important for state environmental professionals to stay abreast of all changes in regulations, policies, and technologies related to their field.

In the role of grant administrator, state environmental officials perform a variety of project management tasks. These often include: conducting seminars for local officials to raise awareness about program opportunities; assisting in conceptualizing local projects which meet the grant criteria; soliciting grant applications; evaluating applications and selecting projects for funding; negotiating contracts, budgets, and schedules; monitoring work progress; conducting site inspections if appropriate; and closing out contracts. There is a current trend of more grant funding and authority being delegated to local governments as they develop increasing technical capabilities. Therefore, it appears that grant administration will become an increasingly important role for state agencies.

State natural resource agencies, like their federal counterparts, manage and develop lands and facilities for recreation (state parks, boat launches, etc.), wildlife management (state game areas and wildlife refuges), and forest management (state forests). Natural resource professionals who staff these sites have occasional opportunities to work in the outdoors as part of their responsibilities (see Chapter 7).

## The "Pros"

The several advantages of state government employment are in many ways similar to those offered by federal agencies. In terms of the work itself, the opportunities to have major influence on public policy, to gain substantial responsibility early in a career, to establish an extensive network of environmental contacts with federal, state, and local officials, interest groups, and industry representatives are basically the same. Likewise, medical and dental benefits, retirement programs, vacation and holidays, job stability, civil service protection (although this may vary from state to state), and professional training and development opportunities are essentially the same for both federal and state employers, with a few exceptions in states where benefits are not as complete. In general, pay is often moderately less at the state level than the federal level, and sometimes much less.

State employment of environmental professionals also offers a variety of employment locations. Besides the headquarters operations, usually located in the same city as the state capitol, positions are also available in sub-state regional and district offices around the state. The work in the headquarters offices tends to be more administrative in nature, while an agency's field work is largely handled by the district offices. At headquarters, the work deals with policy and program development, data collection, major permitting decisions, and program budgeting and resource allocation. It is the field work which involves the most opportunity for contact with the public, for inspecting facilities and operations, and for occasions to work in the out-of-doors.

Environmental professionals employed by state agencies may get the best overall perspective of complex environmental protection programs. Having responsibility for taking federal and state legislative mandates and implementing the required programs, state officials must put concepts into action at the grassroots level. Thus, they are exposed to and affected by political processes of the U.S. Congress and state legislatures, and also gain first hand experience in attempting to manifest those policies in communities and on industrial sites within their jurisdictions. Implementation difficulties and needs for program modifications are communicated back up through the federal system via these same state officials. For environmental protection specialists, at least, state employment offers a unique opportunity to become an environmental program "expert".

## The "Cons"

On the down side, state government work is subject to many of the same political and bureaucratic frustrations as federal employment. The governor's office, legislature, lobbying organizations, and other key players in the state level political arena often have great influence in important environmental and natural resources decisions, totally beyond the control of the state agency official.

State employees must be aware of the "pigeon hole syndrome" which often occurs in bureaucracies. By staying in one position for too many years, individuals may be seen as a specialist in one narrow area, with limited capability to work effectively in other programs. To prevent being "pigeon holed", the author recommends a strategy that has been successful for him. Work in a specific program area for only three to four years as a maximum, then look for new challenges within the agency, within other state government agencies, or even outside of government. Following this strategy will increase the state employees' versatility by demonstrating an ability to adapt to new ideas and situations.

One truth of government work apparent in recent years, particularly in environmental protection programs, is that there are never enough resources to do the job (by personal standards, or according to public demands). State officials must learn to live with fiscal restraints by targeting scarce resources into projects which achieve the most public good.

Because state bureaucrats are much more easily accessible to the public than their federal counterparts, they are more often the target of public outrage and news media scrutiny. Environmental professionals in state government may have their competence questioned, sometimes even in public forums! They need to learn not to take such verbal attacks personally. In balancing interests in making tough decisions, not everyone is going to be happy with the outcome. Industrial representatives usually argue that decisions are too restrictive, while environmental interest groups and citizen organizations charge that decisions are too lenient. Dealing effectively with controversy and crisis situations is, and will continue to be, a challenge for state environmental managers.

A final caution for those desiring to work for state agencies, programs and projects do not always get implemented in the ideal way that they were conceived. Both municipalities and industries, if they do not see an advantage in responding to requests from state officials, will often be uncooperative, and at times even defiant of state authority. Prospective employees need to be prepared to accept the less than perfect art of government program administration as a fact of life within a political environment.

### *LOCAL GOVERNMENT*

The most unique aspect of environmental work at the municipal level, whether it be as a wastewater treatment plant supervisor, a parks and recreation director, an environmental sanitarian, or a city planner, is that it is "grassroots", "hands-on", "on-site", and "face-to-face". Employees of local agencies have the most opportunity for direct and frequent interaction with the public on a day-to-day basis. At times, such interactions may reach high levels of emotional intensity, in the negative sense. This is where government is most accountable. However, repeated contact with individual citizens over time also allows local officials to develop long-term relationships built on trust and mutual respect.

Oftentimes, important environmental issues and controversies begin as a result of a local citizen's complaints to a local agency, or as a result of an inspection or discovery by a local agency. Many environmental professionals see this as "where the action is".

Municipal agencies (including city, county, township, sub-state regional, and metropolitan agencies) range in sophistication from having no environmental staff and deferring all actions to the state agency, to very large professional environmental staffs equal in capability to state and federal agencies. However, by far, most local agencies have relatively small staffs. As a result, environmental professionals in local government tend to be generalists rather than specialists.

Local environmental officials serve the public at the pleasure of a legislative body, such as a city council or county board of commissioners, and sometimes in addition, a special board or commission, such as a planning commission or board of health. Similar to their federal and state cohorts, local government employees get involved in policy development, the enforcement of regulations, citizen complaint handling, and a wide variety of administrative tasks (budgeting, scheduling, reporting, and contract management).

## THE "PROS"

Working as a local government official provides an excellent perspective on the implementation of environmental protection programs. There are usually good opportunities to have a major influence on local public policy decisions, as well as to provide occasional input into state and federal policy development. Because the line of accountability to the public is shortest at the local level, many individuals feel that they can more effectively serve the public interest. Financial resources are often scarce, leaving plenty of room for personal creativity and innovation to successfully implement projects and programs. Finally, the wide variety of work that local environmental officials are asked to perform presents stimulating new challenges on an almost daily basis.

## THE "CONS"

There are a number of limitations and cautions associated with municipal level employment of which environmental professionals should be aware. The image of the overworked, underpaid local official rings true in many agencies. For anyone who has worked in such positions, it seems that there is never enough time or manpower resources to complete important tasks. In addition, except for certain higher level management positions, local government pay scales are often significantly lower than that of state and federal agencies.

Performing environmental work within local government carries the liability of being susceptible to political interference in the form of budget cuts or overturned decisions. Aside from this occasional frustration, political actions of a local legislative body can create job instability or uncertainty.

Lastly, local environmental officials are the first "line of complaint" for the public, and often the easiest target for outraged citizens. Besides feeling somewhat unappreciated in these situations, local bureaucrats are often frustrated to learn that they only have very limited authority to respond to environmental complaints. When local officials do have the authority to respond to incidents, they often find themselves on call 24 hours a day.

## INDUSTRY

Nearly all types of manufacturing industries hire environmental professionals. The "heavy industries" (e.g., steel, chemicals, pulp and paper, power, petroleum, mining, and automobiles), because they have to manage large volumes of solid waste and wastewater and control air emissions, hire substantial numbers of environmental engineers, air scientists, hazardous waste specialists, water quality specialists, hydrogeologists, regulatory compliance specialists, and environmental health and safety professionals. Environmental work within the industrial arena generally takes place at two levels, the plant level and the corporate or headquarters level.

### In-Plant Employment

Environmental specialists working at industrial production facilities focus primarily on: (1) keeping pollutants from being released into the air, water, ground water, or land; and (2) managing the collection, treatment, transport, and disposal of solid and hazardous waste materials. This is the "hands-on" level for environmental management, attempting to keep production processes within the legal requirements of various permits. In-plant environmental control is essentially "in the trenches" problem solving (for things such as equipment breakdowns, chemical spills, training of hourly workers on proper waste handling techniques, wastewater treatment plant flooding, responding to regulatory audits by governmental agencies, adjusting waste treatment systems to achieve operating efficiencies, identifying opportunities for waste minimization, etc.) requiring a high degree of personal resourcefulness and an ability to facilitate cooperation among a wide variety of plant personnel. Because at most industrial plants there is only one, or at most a few, dedicated environmental positions, those that fill the positions become "do everything" generalists, being called upon to respond to any happening that in any way might be considered "environmental".

## The "Pros"

There are several advantages to plant level environmental work. First of all, this work offers the opportunity to address pollution control at the source. Environmental specialists can see the direct results of their efforts. They also have the opportunity to work across all media areas (i.e., air, water, ground water, and land). These professionals perhaps develop the best understanding of how environmental laws and regulations affect industrial production processes. Because in-plant environmental specialists work with a single or specific group of industrial processes, they quickly become experts in the regulations, control technologies, and other environmental aspects associated with those particular processes. In the personal compensation area, in-plant environmental specialists enjoy good pay, usually comprehensive medical and dental benefits, the standard number of vacation days, and other benefits offered by these usually large industrial employers.

## The "Cons"

There are also some limitations to in-plant environmental work. Because industrial production facilities exist to manufacture products (and make money), expenditures for environmental protection or compliance purposes are often seen as a necessary evil. As a result, budget struggles between environmental specialists and plant and corporate managers are common. As in most types of environmental work, there are almost never enough resources to successfully complete a job in the best way possible. To assist environmental personnel with carrying out various types of projects (ranging from feasibility studies to engineering design and construction), companies are increasingly retaining the services of outside consultants and contractors. The result is that in-plant environmental specialists are devoting more of their time to the management and oversight of these contractors. Another sometimes frustrating aspect of in-plant work is the increasing amount of required paperwork. Air and water monitoring reports, hazardous waste manifests, hazardous chemical inventory reports, toxic emissions calculations, laboratory chain-of-custody documents, budgets, schedules, requests for proposals, bid documents, purchase orders, and emergency response plans, are all examples of paperwork which must be completed and/or updated on a periodic basis.

## *Corporate Employment*

Professional environmental work at the corporate level of a company differs significantly from in-plant work. Activities tend to reflect the "big picture" in problem solving, rather than implementation of specific plant level technical solutions. The work is more managerial and administrative in nature, and revolves around coordination, communication, and facilitating cooperation. Corporate environmental staffs can play several roles, including: providing technical assistance to individual plants; taking the lead in sensitive negotiations with governmental regulators; conducting in-house regulatory audits of plants; determining the impact of proposed laws and regulations on the corporation; advising production managers, lawyers, and public relations personnel on environmental matters; and training of both corporate and plant personnel on changing environmental regulations and technologies. In addition to these coordination tasks, corporate environmental staffs also often become involved in the management of large-scale environmental projects (such as the design and construction of a new industrial wastewater treatment facility or hazardous waste incinerator). In this role, corporate officials are often responsible for the selection of qualified contractors, the negotiation of contracts, the monitoring of budgets, schedules, and work progress, construction oversight, and overall quality control. Headquarters environmental staff also get involved in the siting of new and expanding industrial facilities. Determining the environmental impacts and permitting requirements of new operations are critical considerations prior to such corporate investment decisions. Most corporations prefer to bring environmental specialists onto the corporate staff after several years of experience at the plant level.

### The "Pros"

The advantages of corporate environmental work include the opportunity to specialize in a specific program area, to have substantial influence on the operations and development of a multiple of industrial facilities, and to stay on the "cutting edge" of regulations and technology. Personal compensation for corporate staffs include good pay, medical and dental benefits, vacation, retirement programs, and usually some type of stock option or other investment mechanism.

## The "Cons"

There are relatively few limitations to corporate environmental employment. However, some corporate managers feel that over a number of years they begin to lose touch with the technical and "hands-on" aspects of work within the plants. With some corporations, headquarters staff are expected to travel extensively to plants across the country, and sometimes around the world. Where corporations have multiple division and regional corporate offices, environmental managers may be required to relocate to such offices as corporate needs change or as promotions occur.

Special note: The author recommends that individuals drawn towards industrial environmental work seek out employers who have progressive, or at least responsible, attitudes or philosophies towards corporate obligations for environmental protection. This will reduce a potential source of frustration for environmental professionals trying to do the "right thing".

### *Non-Manufacturing*

Several types of non-manufacturing businesses also hire environmental professionals, usually for very specific purposes. For example, commercial banks and other lenders may employ environmental specialists to conduct or coordinate environmental site assessments before foreclosure or lending on commercial real estate. Insurance companies may use environmental specialists to review the regulatory compliance records of potential clients, to check whether a company has a workable chemical emergency response plan, or is otherwise subject to significant environmental liabilities. Another example is the computer software company that employs various types of environmental professionals to develop specific computer applications or data bases for environmental projects or programs. Hazardous waste transportation companies hire regulatory specialists to ensure manifests are properly prepared and submitted to the appropriate agencies, and to conduct compliance audits of disposal facilities. Because each of these unrelated types of businesses has a unique reason for employing environmental professionals, there are far too many to practically introduce in this discussion. Aspiring environmental professionals need to understand that employment options exists for both the traditional heavy manufacturing industries, as well as for service oriented businesses. However, personal creativity and initiative are needed to identify companies in the latter group.

## CONSULTING FIRMS

Environmental consulting firms may be the largest employer group for environmental professionals nationally. These companies consist primarily of scientists, engineers, and technicians whose services are sold to businesses and government agencies. Some of the larger national firms have over 1500 employees. They provide technical capabilities to organizations, usually on a contract basis, who cannot afford, or do not have the need, to maintain a large environmental staff.

Industries and governments hire consulting firms to perform special projects, as well as to assist with day-to-day operations. Examples of special projects might include: conducting an investigation of soil and ground water at a site where contamination is suspected; collecting air samples and loading the results into a computer model to determine environmental impacts; evaluating alternative ways to treat and dispose of a particular type of hazardous waste; and designing a wastewater treatment facility or a leachate collection system for a solid waste landfill. Consultants are also retained to perform more routine environmental work, such as conducting periodic regulatory audits of industrial plants, providing engineering solutions for troublesome process operations which result in accidental releases of pollutants, and performing analyses on complex regulations to determine the most effective strategy for a client to follow.

The scientists and engineers who work as managers for consulting companies generally must perform a variety of business and project management functions aside from their technical work. These functions include sales and marketing, conceptualizing and organizing project tasks, projecting the work effort required for various tasks, estimating project costs, negotiating contracts, supervising project teams, monitoring work progress, maintaining client relations, and verifying the successful completion of work.

In relation to other environmental employers, the work in consulting companies has some special characteristics. A fast-paced working environment is common among these firms. On any given day, the consultant's activities might include drafting a project proposal, overseeing field work, attending a marketing meeting, researching a regulation, writing a project report, calling a client to resolve a budget issue, and recruiting a scientist or engineer. Generally, consultants undertake a much wider variety of technical assignments than their environmental professional counterparts in government and industry. This is because consultants work for many different industries and government agencies, each with their own unique set of environmental needs. Professional credentials are probably more important for individuals working in consulting firms than for persons working for any other type of employer. Because professional credentials help project the appearance of competence, consulting firms prefer to hire people with master's or Ph.D degrees, Professional Engineer certification, Professional Geologist certification, and other more specific certifications, such as for asbestos contractors, hazardous waste managers, site assessment specialists, etc.

Another employer of environmental professionals, similar to consulting firms, are the architecture-engineering firms (commonly referred to as "A&E" firms). These companies provide conceptual and detailed architectural and engineering design services, often along with construction management services, for buildings, industrial plants, and associated facilities. Wastewater treatment plants, waste incinerators, landfills, hazardous waste storage buildings, tank farms, and other environmental management facilities are designed and built by A&E companies. Environmental engineers, scientists, and regulatory specialists are commonly employed by these companies, primarily to perform the permitting required for the projects and to determine what specific design criteria are required by environmental regulations.

## THE "PROS"

Environmental work within a consulting firm generally offers four primary advantages. First, consultants are usually exposed to more differing types of interesting technical work, and more projects, within a given period of time than environmental professionals working for other employers. Second, the quick pace of the work environment can be very stimulating and challenging, especially for "high-energy" individuals. Boredom is rare. Third, because of the generally strong demand for consulting services, and the scarcity of qualified engineers and scientists, consulting firms usually offer very good pay and benefits. And fourth, depending on the size, rate of growth, and profitability of a particular consulting firm, there are often excellent opportunities for advancement and promotions. As a person's value to the company becomes increasingly evident over time, pay and "perks" increase accordingly.

The potential rewards of consulting are so lucrative for many entrepreneurial individuals that many, after several years of experience, try consulting on their own. The ones that have the easiest time succeeding are those who have found or developed a "niche" area of expertise and potential client contacts.

### THE "CONS"

Along with the possible rewards of consulting there are a few drawbacks. That fast-paced, often hectic, work environment which provides challenge and stimulation, can also produce substantial stress. Attempting to meet deadlines, talk with clients, coordinate work, make sales calls, write reports, and perform similar tasks can become overwhelming at times. An associated concern is that when the workload is heavy and the deadlines short, consulting can require long hours in the office. At many national and international consulting companies, clients can be located nearly anywhere in the country, or even the world. It is often expected that employees be willing to travel if required by the needs of a project or client. A large number of environmental consultants must travel extensively, which can make things difficult for those with families or community commitments. Lastly, the demand for environmental consulting services fluctuates with the economy, and in recessionary times job stability can become an issue of concern. Layoffs are more likely to take place within consulting companies that operate "close to the margin", thus being unable to carry individuals without billable (to a client) work on overhead budgets for a very long period of time.

## ENVIRONMENTAL ORGANIZATIONS

Many environmental professionals have found rewarding careers with various environmental organizations. These include environmental interest groups, educational organizations, and research institutes, each of which has some specified mission or philosophy. These groups usually focus on educational activities and public awareness building, and sometimes get involved in direct lobbying activities.

Much of the work activity at environmental organizations is communications related. Professional staffers get involved in: organizing mass mailings; writing press releases; fundraising; conducting educational seminars; telephone canvassing; coordinating opinion surveys; producing brochures and flyers; and presenting testimony at public hearings or legislative committee meetings. In addition, they usually provide staff services to a board of directors, or similar group, where assignments might include: reviewing and analyzing proposed legislation; researching and preparing position papers on specific environmental topics; and providing administrative support (setting meeting dates, determining meeting locations, taking minutes of meetings, following by-laws in performing procedural tasks, identifying speakers, etc.).

To profile the National Audubon Society as an example, this organization has 293 full-time employees with position titles like: Educational Coordinator, Editor, Public Lands Specialist, Camp Director, Staff Scientist, Sanctuary Manager, Research Biologist, Environmental Policy Analyst, Research Biologist, Warden, and Volunteer Coordinator. The efforts of the professional staff are further enhanced by a strong grassroots network of volunteers coordinated through local chapters. The National Aububon Society is one of the largest of the environmental nonprofit organizations in the U.S. with over half a million members.

The strength of the several national environmental organizations comes from their membership. Another aspect of professional work with these types of employers is recruiting new members and providing member services. The latter often includes producing a newsletter or other type of periodical, in some cases a full-color "slick and glossy" magazine. Also, finding and delivering "perks" (tours packages, rental car discounts, unique artwork, etc.) to members can be important. The overall influence and stability of the organization may depend on the satisfaction of its members.

Each environmental organization, interest group, or research institute has its own particular mission from which it derives its reason for existence and a rationale for its activities. For example, The Nature Conservancy is dedicated to the identification and protection of ecologically important lands. The National Wildlife Federation and the Audubon Society are citizen conservation education organizations that have their roots in wildlife and natural resources management issues. The Natural Resources Defense Council is focused on environmental and natural resource protection through scientific research and legal means. Beyond these national organizations, there are countless state and local groups that have as their mission a specific environmental issue, such as promoting recycling, cleaning up a contaminated industrial site, campaigning for the purchase and development of land for a new park, or raising public awareness of water pollution along a local waterway.

One important aspect about environmental organizations is that these groups offer excellent opportunities for volunteers and student interns desiring to work in the environmental field. They provide future environmental professionals a chance to get a "feel" for organizational work and to see the "big picture" perspective on specific issues. The careers of many environmental specialists began with volunteer or internship work with a nonprofit group.

## THE "PROS"

There are several positive aspects to professional work with environmental organizations. Most of these organizations pride themselves on having significant influence on environmental programs, policies, and actions by government or industry. At the national level, this may mean affecting an action of a federal agency to protect a park. At the local level, it may mean having a local plant change its operating procedures to reduce noxious odors. In addition, the nature of the work allows professionals to "network" extensively with government officials, politicians, industrial representatives, the news media, and other interested organizations. And because the staffs tend to be small in these environmental groups, professional creativity and innovation is welcomed and encouraged.

### THE "CONS"

The two primary drawbacks of employment with environmental organizations are relatively low pay and limited promotional opportunities. Except at the largest, and best funded, national organizations, professional pay scales are often substantially lower than equivalent positions in government or industry. Furthermore, the small staff size often results in limited advancement potential within the organization. However, moving up the career ladder between these organizations may in essence be a promotion.

As a final note, some of these organizations (i.e., the ones who collect and analyze facts and advocate responsible positions on issues) are held in high regard by government and industry. Other environmental organizations (i.e., those that depend more on human emotions than facts, and take more radical positions on issues) are not always respected by major players in the environmental community, and therefore have little influence on public issues.

## ACADEMIA

Graduates of environmental programs holding or pursuing master's or doctoral degrees often target academia as their choice of employers. Working as faculty on the collegiate level involves primarily teaching and research activities. Among the responsibilities of faculty members teaching courses are participating in curriculum development, reviewing potential class materials (textbooks, articles, etc.), planning and scheduling courses, lecturing, facilitating classroom discussion, organizing field experiences and laboratory activities, preparing and correcting tests and quizzes, arranging for guest speakers, grading term papers and projects, supervising teaching assistants, and counseling students. On the research side, professors perform the following duties: organizing research projects, estimating budgets, preparing research grant applications, supervising research assistants, monitoring project activities, and writing technical papers on the results of projects.

Beyond teaching and research, faculty members get involved in a wide variety of other activities, such as serving on faculty committees, performing administrative tasks, participating in technical seminars, providing testimony in court as an expert witness, being an information source for the news media regarding an environmental controversy, and consulting on special government projects.

Virtually every major four-year college, and most two-year colleges, have environmental (or related) faculty positions. Openings at major universities are relatively scarce and much sought after, making the competition fierce. Many more less competitive opportunities exist with smaller four-year institutions and two-year colleges, especially community colleges. These positions offer faculty members a chance to get experience and build their teaching credentials, to perhaps better compete for more visible and lucrative positions at large universities. To get started on a career path towards academia, students usually serve as teaching or research assistants while completing a graduate level degree or conducting specialized types of research. Through these "assistantships" students can begin building a network of contacts who are able to assist them in finding permanent employment.

## The "Pros"

College level faculty find several advantages to their work in academia. Professors have the opportunity to study and teach in their specific areas of interest, as opposed to most environmental professionals in government and industry. They have a legitimate reason to stay at the "cutting edge" of develop-ments in their fields. Similarly, opportunities to get involved in interesting research projects often exist for faculty members, particularly at the larger colleges and universities. For those who enjoy intensive interaction with motivated and interested students, teaching can provide tremendous intellectual stimulation. Professors also have repeated chances to travel to desired destinations for technical conferences, professional association meetings, and academic conclaves. Summers are usually available to faculty members to pursue special research or teaching opportunities (or to take vacations!). One last advantage to employment in the academic sector is the prestige, and even respect, associated with being a college professor.

### THE "CONS"

Some of the less positive aspects of academic employment include: low to modest salaries; having to work within the political environment of an academic bureaucracy; occasionally being overburdened with administrative responsibilities; and having no direct feedback on how educational or research efforts may have led to actual environmental improvements.

## OTHER EMPLOYERS OF ENVIRONMENTAL PROFESSIONALS

There exist a number of other employers of environmental professionals in addition to the ones discussed in this chapter. However, they generally do not employ large numbers of environmental specialists at a single location. Still, these organizations do provide employment niches for individuals desiring to work in the environmental field. Examples of these employers, many of which have already been mentioned in other chapters, are trade associations, professional associations, banks, insurance companies, public school systems, mass media and communication companies, law firms, and public relations firms.

## REFERENCE

Personal communication with the National Audubon Society, January 16, 1991.

# CHAPTER 11

# A Word About Salaries

# A Word About Salaries

Let's begin our discussion of the sensitive subject of salaries with a few gross generalizations. First of all, most people entering the environmental field do so for reasons other than trying to make the highest salary possible. Second, although these people are motivated largely by nonfinancial reasons, they can and do make a good living, as defined in many ways, including monetarily.

For the most part, there is not a large variance in starting salaries among the majority of environmental careers discussed in this book. Entry-level professional positions usually begin around \$17,000 per year and range upward to around \$30,000. There are exceptions to this statement, of course. For example, a recreation coordinator for a local government agency in a rural area may make only \$15,000, or a newly graduated chemical engineer with a minor in environmental engineering might be hired by an industrial employer for \$32,000 or more. These examples, however, are more the exception than the rule, with the majority of entry-level positions beginning in the low to mid-\$20,000s.

The results of a career survey conducted by Cahners Research for *Pollution Engineering* magazine, appearing in its July 1991 issue, presented several findings of interest to potential environmental professionals. According to the survey of readers, the average annual salary for environmental professionals (regardless of age, experience, and educational level) is \$53,169. The average for those working in manufacturing is \$55,000; for those working at electric or gas utilities the average is \$57,228; and for those in an "other" category (which includes consultants), the average is \$63,246 annually. It was found that the annual average salary for government workers is about \$8500 less than their counterparts in manufacturing, coming in at \$46,472. Also, respondents reported that they work nearly 48 hours per week, regardless of employment sector. Why do environ-mental professionals work these longer hours? Over 90% indicated that they were happy with their jobs. A final important finding of the survey was that 65%

said that a feeling of accomplishment was the most rewarding aspect of their jobs.

Another recent study surveying readers of *Hazmat World* magazine who work for industrial employers (i.e., general manufacturing and processing, chemical, and petroleum industry employers) provides some additional salary figures. The annual average salary for these industrial environmental professionals and technicians is $47,528. By educational levels, the average annual salary breakdown in this survey was as follows: for those having only high school degrees, $38,444; for technical school graduates, $39,636; for managers and scientists with bachelor's degrees, $46,971; for those holding master's degrees, $53,579; and for those with doctorates, $61,800.

The level of compensation for environmental professionals is more often determined by the type of employer, the experience of the employee, and the location of employment, than by the specific degree held or career path chosen. The remainder of this chapter will address these factors as well as others which may influence salary levels.

## GOVERNMENT

Government agencies at the federal, state, and local levels employ large numbers of environmental professionals. In some professions, such as wildlife biology and environmental enforcement, government is the major employer. Overall, salaries of government employees tend to be less than for their counterparts working in the industrial or consulting sectors. However, many people prefer government employment because, although the pay may be in the moderate range, other benefits make up for it. For example, government agencies often offer more complete medical and dental benefits, as well as vacation time and administrative leave than private sector employers. Traditionally, it has been thought that another important benefit of government employment was job stability (this may hold some truth for the federal government, but in recent years several state and local governments have faced hiring freezes and even layoffs due to budget problems, and sometimes politics). Also, many professionals identify closely with the philosophical orientation of government (i.e., working to protect and manage the environment for the public good).

Of the various levels of government, the federal government generally offers the highest salaries. However, many state and metropolitan area local governments are catching up as they become increasingly involved with resource management and environmental protection issues. Local governments in rural areas usually offer the lowest salaries, often due to a limited tax base or to a lower political priority assigned to environmental activities. Almost all federal and state agency salary offerings are administered through a civil service system. Civil service rules tend to keep consistency across the professions represented within government agencies. For example, an aquatic biologist and a policy analyst working at the same level of responsibility would make roughly the same salary. However, some civil service professions, notably engineering and law, may be allowed some salary advantages due to their specific training requirements and the outside competition for candidates.

Civil service positions in the federal government are classified according to GS levels, with GS-3 being the entry-level for technicians without experience, GS-5 for those holding bachelor's degrees, GS-7 for those holding master's degrees, and GS-11 for those with doctoral degrees. Entry-level job candidates with strong academic records and/or relevant experience may be able to get classified at the next highest GS level (i.e., GS-7 for bachelor's, GS-9 for master's, and GS-12 for doctorates). At each GS level there are ten yearly salary step increases allowed. The 1992 GS salary level ranges are:

| | **Step 1 Step 10** | |
|---|---|---|
| GS-3 | $12,531 — $16,293 | $14,082 — $18,303 |
| GS-5 | $15,738 — $20,463 | $17,686 — $22,996 |
| GS-7 | $19,493 — $25,343 | $21,906 — $28,476 |
| GS-9 | $23,846 — $31,001 | $26,798 — $34,835 |
| GS-11 | $28,852 — $37,510 | $32,423 — $42,152 |
| GS-12 | $34,580 — $44,957 | $38,861 — $50,516 |

The higher GS levels, 13 through 15, require substantial experience and have salaries ranging to a maximum of $83,502. A limited number of executive positions are set at the highest 1992 level of $143,800.

State government agencies offer salaries in ranges similar to the federal government's GS system. Taking the field fisheries management as an example, a recent survey of 45 state agencies showed that: the average salary of technicians ranged from $16,099 to $23,788, similar to the federal GS-3 level; the average salary of a junior fisheries biologist ranged from $20,844 to $29,938, similar to the federal GS-5 level; the average salary of a senior fisheries biologist ranged from $23,173 to $34,052, similar to the federal GS-7 level; the average salary of a hatchery manager ranged from $23,184 to $33,532 and that of a division chief ranged from $37,622 to $51,185, both similar to the higher federal GS levels.

## INDUSTRY

Industry hires environmental professionals to work at both the manufacturing plant and corporate level. Salaries for these positions vary depending on the size of the company, the characteristics of the production processes (how much pollution is generated), and the number of people supervised. In general, though, salaries for industrial positions tend to be somewhat higher than comparable ones in government, academia, and nonprofit organizations.

An environmental position in a small, conservative manufacturing company may be seen as a necessary evil and not contributing to the company's profit margin. As such, the salary for that position may be quite modest. On the other hand, a progressively managed company may recognize that an effective environmental staff contributes directly to the company's bottom line, saving it hundreds of thousands of dollars by minimizing environmental liabilities (i.e., avoiding costly lawsuits) and maintaining a positive public image. Environmental professionals working for this type of company may be well compensated for their efforts.

## ACADEMIA

Entry-level research and teaching positions at colleges and universities generally have low salaries, but these can escalate steadily with experience, professional recognition (particularly through the publishing of papers, articles, and books), and through participation in research projects funded with outside grants from government, corporations, or nonprofit organizations. Although assistant and associate professors make only moderate level salaries, those who achieve tenure as full professors or rise to high level administrative positions (i.e., dean, director of a research institute, or curator of a botanical garden) can earn very respectable incomes.

It is important to point out that there may be dramatic differences in compensation among various institutions of higher education. For example, professors at community colleges and private colleges often make only a fraction of what their counterparts in large public universities earn in salary (this is not to say, however, that there are not other substantial benefits and amenities associated with teaching at the smaller colleges).

A 1989 salary survey of wildlife and fishery teaching positions at 43 American universities conducted by the National Wildlife Federation provides a good indication of current academic salary levels. The survey results showed that: full professors earned salaries ranging from $44,886 to $59,715; associate professors earned between $38,487 and $49,675; assistant professors earned from $32,947 to $41,962; and instructors earned from $24,284 to $34,366.

## CONSULTING

Consulting firms that provide natural resources management and environmental protection expertise to government and industry offer a wide range of salaries and salary earning potential. Although entry-level salaries may be similar to equivalent government or industry positions, the opportunities for rapid salary increases are good, especially as a person's contribution to the company becomes known. This is particularly true for consulting firms working in the fields of hazardous waste management, contaminated site cleanups, air and water pollution control, and regulatory compliance. For experienced environmental professionals, consulting firms may offer the highest salaries available, particularly when bonuses, profit sharing, and other compensations are included.

## ENVIRONMENTAL ORGANIZATIONS

Salaries for professional positions with environmental organizations range from those as high as state government positions, within the larger national organizations, to ones much lower within state/regional/local groups or specific issue-oriented organizations. For example, at the National Audubon Society, professional salaries begin around $13,500 and range to over $100,000 at the highest executive levels. On the other hand, the much smaller local or issue-oriented nonprofit organizations may offer only an hourly wage of $4 to $5, or less. And many of these organizations have numerous unpaid volunteer positions available.

## GEOGRAPHICAL DIFFERENCES

Salaries for environmental professionals also vary from region to region around the country. Such differences usually correlate directly with the overall cost of living in a region (and competitiveness for candidates). Salaries tend to be higher in the northeast, the mid-Atlantic coast, and the west coast. They tend to be more moderate in the midwest, southeast, and northwest. Salaries for similar positions are usually the lowest in the deep south and the western mountain states. Job seekers should expect these regional variations when deciding where to work.

Variations in salaries also occur between urban and rural areas, even within the same region. Urban areas have a higher cost of living which is reflected in salary levels. The converse is often true for rural areas. Perhaps the most dramatic example of these differences is with local governments. City and county agencies located in large metropolitan areas can draw on a substantial and somewhat diverse tax base to generate revenue to pay employees. The salaries for these employees are higher than those of their rural counterparts, to compensate for higher real estate costs and (ironically) taxes. A rural county government, on the other hand, may have only a few manufacturing plants and a handful of small businesses from which to generate tax revenue. This situation, coupled with a lower cost of living, results in lower salaries for county employees (however, for many people the benefits of small town and country living more than make up for the salary differential). This urban-rural salary difference is not as prevalent with federal and state government agencies, nor is it with national scale industrial employers.

## THE VALUE OF EXPERIENCE

Experience is perhaps the most influential factor in determining the salary level for an environmental professional. The simple fact is that the more professional experience a person has, the more valuable he or she is to an organization. Even at the entry-level positions, experience is critical (this will be discussed in detail in Chapter 13).

In government, position descriptions and salary levels are established by civil service criteria. Although government managers hiring new employees have some discretion in placing individuals within the salary structure, there is little room for negotiation. This is generally not true with other types of employers who have less bureaucracy to deal with, such as consulting firms, nonprofit organizations, and industry. With these employers, there is often more opportunity for job candidates to negotiate salary levels, except at the entry level.

As employees gain professional experience, they gain increasing leverage in salary negotiations with their employers, or prospective employers. There seems to be three levels of experience for the employee, where negotiating clout clearly increases:

- 0 to 5 years — set entry-level salary levels; usually technical positions
- 6 to 15 years — some moderate room to negotiate; senior technical and supervisor positions
- 15 plus years — maximum negotiating clout; senior level management or technical positions

Most of the employee shortages within the environmental professions are for those positions requiring ten or more years of working experience. Because of this scarcity, qualified and experienced candidates for certain high demand positions can essentially "write their own ticket".

## EDUCATIONAL LEVEL

Most people assume that an employee's salary would be commensurate with the level of education attained. This is basically true for the environmental professions, particularly for entry-level positions. Technicians and field specialists without bachelor's degrees generally earn lower salaries than those holding bachelor's degrees. Likewise, those with master's degrees often have some salary advantage over those with only a bachelor's degree. In the case of a doctorate degree, the picture is not as clear.

For research and college level teaching positions, a Ph.D. has a definite salary advantage because it is the educational standard for research institutions and universities. Also, for certain high visibility administrative positions in government and the nonprofit sector, and for particular management positions in consulting firms where such credentials are held in high regard, a doctoral degree can yield additional pay and compensation. However, most environmental positions within government, industry, and the nonprofit sector do not require a doctorate. In those positions, individuals holding a Ph.D. often do not make more than those with only a master's degree, or in some cases, than an experienced person with only a bachelor's degree. In fact, oftentimes a job candidate who holds a doctorate for a position which requires less educational background, is seen as overqualified and likely to become unhappy with duties assigned to that position.

The above discussion focused on the salary differences stemming from level of educational background. Once an individual has specific working experience such generalizations become much weaker. Based on an individual's seniority with a single employer and/or his or her value to an organization (i.e., performance), a person with less formal education but with more practical experience than a coworker can earn a significantly higher salary than the coworker. This situation happens more often with employers who are not bound by bureaucratic procedures and systems.

## WHO MAKES THE MOST?

In the author's opinion, there are two types of environmental professionals that are able to draw the highest levels of salary. First are the top managers, presidents, and chief executive officers of industry and government. Second are the risk-taking environmental entrepreneurs.

The environmental professionals who attain high level management positions in large organizations are those who probably deliberately chose a management career path and worked to develop sound organizational and management skills. For these individuals, their environmental specialty has become less important than their ability to provide leadership, communicate, estimate budgets, develop policy, and perform related management tasks. Such top management positions exist in nearly every area of environmental work, whether it be the director of a state fish and game agency or the chief executive officer of a large hazardous waste treatment and disposal company. The salaries for these managers often fall within the $80,000 to $120,000 range, and may go higher. However, the number of high level management positions in government and business is limited, and generally only the best and most experienced performers are appointed to these offices.

Environmental entrepreneurs, on the other hand, are those risk-taking individuals who start up business enterprises that are in some way related to natural resources or the environment. Examples of such businesses are consulting firms; companies that manufacture or distribute pollution control equipment; environmental laboratories; companies that develop computer software for environmental or natural resource management applications; and field equipment suppliers. One enterprising wildlife biologist (an acquaintance of the author) who had trouble landing a government job, established a wild bird store which sells seed and seed mixtures, bird guides, binoculars, bird feeders and houses, bird song recordings, and related items.

Merely the fact that someone is an entrepreneur does not imply that he or she makes one of the highest salaries in the environmental field. In fact, many aspiring entrepreneurs start businesses that fail within the first year or two. Others step into "entrepreneurship" only as a part-time endeavor, or as a means of financing their recreational or avocational interests (i.e., a charter boat operator who conducts nature tours, or a retired teacher who is writing an environmental education textbook). But, for those entrepreneurs who dedicate themselves to a niche in the environmental marketplace, their success may provide financial rewards substantially exceeding those of an environmental professional employed by a government agency, academia, or a business organization.

Such financial success does not happen overnight. Oftentimes these business people make nothing or relatively low salaries during the business start-up period. Only after a period of sustained effort can personal salaries begin to exceed those of employees of a large organization. Although for the entrepreneur the amount of compensation is not limited by civil service guidelines or corporate pay scales, to be successful, sound business skills and judgment are required. This is, of course, beyond whatever environmental training or education is needed. Despite the sacrifices necessary to become a successful entrepreneur, increasing numbers of environmental professionals are steering their careers in this direction.

## BENEFITS BEYOND SALARY: BOTH TANGIBLE AND INTANGIBLE

Salary is only one dimension of the total compensation picture for an environmental professional. Beyond salary, other tangible employee benefits (ones which can be assigned a dollar value) may include medical insurance, dental insurance, disability compensation, group life insurance, profit sharing programs (such as an ESOP — Employee Stock Ownership Program), administrative leave, a retirement program (pension, IRA, Keogh Plan, etc.), subsidized day care for children, dependent care, and tax deferred savings or investments.

In addition to direct benefits offered by an employer, there are a variety of other factors upon which it is difficult or impossible to place a dollar value, but which might also be construed as compensation. These include: a positive physical and social working environment; flexible working hours; easy access to recreational areas and facilities; a strongly desired rural or urban lifestyle; a pleasing climate; and perhaps most important, the opportunity to perform meaningful work for the good of society.

It is often these intangible "benefits" which motivate environmental and natural resource professionals to accept particular jobs that do not offer the highest possible salary levels.

## REFERENCES

Bishop, J., 1990 salary survey of hazmat managers, *Hazmat World*, December 1990, vol. 3, no. 12, Tower-Borner Publishing, Inc., Glen Ellyn, Illinois, p. 36-41.

Pirocanac, D., How does your salary compare?, *Pollution Engineering*, July 1991, vol. 23, no. 7, The Cahners Publishing Company, Newton, Massachusetts, p. 12-15.

National Wildlife Federation, *A Survey of Compensation in the Fields of Fish and Wildlife Management*, 25th Edition, National Wildlife Federation, Washington, D.C., 1989.

Personal communication with the National Audubon Society, January 16, 1991.

Federal Employees' News Digest, November 18, 1991, vol. 41, no. 16, Copyright 1991 by Federal Employees' News Digest, Inc. (ISSN. 0430-1692).

CHAPTER 12

# Mid-Career Changes

# Mid-Career Changes

For those people seeking to change careers after several years of professional employment (or unemployment), the environmental field offers tremendous opportunities. Even if a particular individual is employed in a totally unrelated field, there are paths to follow which offer meaningful work related to the protection of the environment and the management of natural resources. This chapter will explore effective strategies to identify environmental career options, as well as to provide some advice on how to maximize one' chances of successfully landing a desired position.

Before attempting to identify a specific career option to target, career changers need to take the time to assess their current knowledge, skills, and interests. There are several career assessment guides available to assist individuals in determining the type of work they would enjoy and to identify their personal strengths (one such book found useful by the author is *What Color Is Your Parachute?* by Richard Nelson Bolles, published by Ten Speed Press, Berkley, CA, 1984). Also, career counselors are equipped with a variety of aptitude evaluation tests to help sort out personal abilities and preferences. The purpose of this personal assessment exercise is to build upon a person's current background, experience, and maturity. Career changers do not need to start from scratch.

Once personal aptitudes and interests are determined, the next step is to relate them to a specific type of environmental work. By targeting a specific career path, the career changer increases the chances of making a satisfying transition. Ideally, the career target should be one which is a blend of a person's subject area of interest and personal skills and knowledge. For example, a professional accountant with strengths in economics and mathematics may have an active interest in finding practical solutions to municipal solid waste problems. Such a person might target a position which is involved in evaluating the economic feasibility of solid waste management alternatives (perhaps comparing the costs of incinerators versus recycling programs) within a government agency or consulting firm.

## BUILDING ENVIRONMENTAL CREDENTIALS WITHOUT ADDITIONAL FORMAL EDUCATION

The smoothest career transitions into the environmental field are those which do not require extensive additional formal education. In the preceding example, the accountant can take several steps to build credentials in the area of municipal solid waste management:

**Read voraciously** — Anyone entering a new career field needs to read as much as possible about what is going on in that field. It is particularly useful to read not just current newspaper and magazine articles, but also the key technical and trade journals. Such journals will bring the reader up-to-date on current issues within the profession or industry. In addition, for areas where the prospective career changer has gaps in personal knowledge, especially in technical areas, it may be useful to read selected textbooks recommended by a practicing professional. Being able to speak and understand the technical terminology of a specific environmental discipline is the first step for an aspiring professional.

**Enroll in training seminars** — Federal and state government agencies, professional associations, universities, and industrial groups all occasionally offer educational seminars on specific environmental protection and natural resource management topics. Attending such seminars is an excellent way to become familiar with the regulations, techniques, tools, technologies, and issues associated with a particular environmental discipline. Also, these seminars often present good opportunities to meet practicing professionals.

**Volunteer** — Experience is always a critical factor for employers in hiring. One way to get direct experience in a specific environmental field (even while still working in a different field) is to volunteer at an agency or organization. Most national environmental organizations and most state and local agencies provide some type of opportunities for volunteers. Although the actual work of the volunteer may occasionally be routine or mundane, the experience will allow the volunteer to become familiar with the organization and the issues which it addresses. Volunteering often also provides opportunities to establish contacts with a variety of environmental professionals and employers.

**Write articles** — Using the environmental background developed through reading, attending training seminars, and volunteering, career changers should write articles in their subject field of interest. The purpose of writing is to associate the career changer's name with the selected specialty area. These articles do not need to reach the sophistication of scientific research papers, but rather can be project updates, reports on new regulations or technologies, or even news pieces about agencies and organizations. Association newsletters, trade magazines, government publications, and nonprofit organization bulletins all accept articles on topics relevant to their readership. The trick is to identify which publications may be read by a desired potential employer.

**Be a public speaker** — This should not be as intimidating as it sounds, particularly for those who enjoy public speaking. Simply research some issue of professional or personal interest, develop a new "slant" on the information (perhaps initiate a survey of individuals embroiled in the issue, or conduct an in-depth interview of leaders in a technical field relevant to the issue), and organize the information in an interesting way for a presentation. Preferably, the presentation should be to professionals or managers working in the area of the speaker's personal interest. The subject of the presentation need not be technical, but it should be of interest to the intended target audience.

**Join and participate in professional associations —** Professionals in nearly every walk of life have at one time or another belonged to at least one professional association. The primary advantages of membership in these groups are to have forums to discuss critical issues, to stay up-to-date on changes affecting the profession, and to associate with one's peers. For the career changer, joining professional associations is a must (what better group is there to inform of a career changer's intentions?). Beyond just membership, prospective professionals should take full advantage of an association's opportunities by actively participating on a technical committee, volunteering to produce a newsletter for a year, or organizing the next annual conference (be sure to get the task of lining up the speakers!).

**Join and participate in trade associations** — Depending on the specific type of environmental work targeted by the career changer, membership and/or participation in a trade association may provide an additional source of employment contacts. There are industry associations representing chemical, petroleum, pulp and paper, printing, auto manufacturing, steel, electroplating, electronics, textiles, mining, power and water utilities, and a host of other industries. If employment in a specific geographic region is desired, it would be wise to identify the types of major industrial employers in the area and their respective associations. Career changers should participate in trade and business groups in the same ways they would in professional associations discussed earlier. Also, the local, regional, or state chambers of commerce might offer similar opportunities for participation and networking.

**Develop a network of practicing professionals** — By taking the initiative and performing the tasks presented, a pool of professional contacts will be developed. Mid-career changers must maximize the advantages of this network of potential employers by: (1) informing all acquaintances about one's personal interest and intentions in obtaining employment in the environmental field; and (2) talking with each contact on a regular basis about possible opportunities and advice. It is critical that job changers not let this valuable resource fall apart from neglect!

## THE BACKDOOR STRATEGY

One strategy that has worked well for a number of mid-career professionals is to accept a non-environmental position with an employer, usually in government or industry, who also offers environmental positions of the type ultimately desired by the applicant. Perhaps an individual's best chance to work for an agency or company is as a human resources manager, even though he/she hopes to land a position involved in cleaning abandoned hazardous waste sites. Using the backdoor strategy, this person would accept the human resources position with the desired employer in good faith and perform to the best of his/her ability. During this employment period, the employee learns about the organization, about the types of positions involved in hazardous waste cleanup, and about the minimum requirements for those positions. While building seniority and a strong work record with the employer, the human resources manager is building the credentials for the hazardous waste position by taking college level science, math, and engineering courses and by performing the steps discussed in the previous section. After an appropriate period of time (perhaps 6 months) the human resources manager can quietly begin to open up a dialogue with the hazardous waste cleanup managers about the possibility of transferring into an environmental position. The entire process, if successful (which it often is!), may take 1 to 3 years, so patience and persistence are absolutely required for this endeavor.

## WHEN ADDITIONAL FORMAL EDUCATION IS REQUIRED

The more dramatic career transitions, such as from a poet to an environmental engineer (to use an extreme example), will definitely require additional formal college level education. However, in many cases it is not necessary to earn a complete bachelor's or master's degree in a field of study. Rather, additional selected coursework, usually in science, math, and/or engineering, may adequately fill the gaps in a person's knowledge base. This selected coursework, combined with other relevant experience and credentials, may fulfill the minimum requirements for several types of environmental positions.

Some environmental careers do require a specific type of degree, often from schools with academic programs of study certified by a professional association, and/or state registration which requires the successful completion of written and oral exams. Examples of these careers include lawyers, engineers, landscape architects, some types of geologists, and industrial hygienists. Before embarking on a specific degree program, aspiring environmental career changers must clearly understand the minimum degree requirements of a targeted profession. When additional formal education is absolutely necessary, the programs at several institutions of higher education should be evaluated, and the one that most directly meets the requirements of the desired position should be selected. Often, however, the career changer must continue working while attending college part-time. In these cases, geographic and academic convenience usually become key factors in choosing a school. Where time permits, those involved in career transitions should attempt to build other credentials beyond just college courses (i.e., volunteering, participating in professional societies, etc.).

## PROFESSIONAL CERTIFICATIONS

Several environmental professional organizations offer certification programs to recognize an individual's expertise and/or experience in specific types of environmental work or procedures. Programs are available in hazardous waste management, wetlands delineation, environmental property assessments, environmental training, industrial hygiene, laboratory technology, and a wide variety of other environmental related disciplines. Generally, specific types of academic degrees are not prerequisites for certifications. Actual experience, at least one or two years at an absolute minimum, and the passage of a written and/or oral examination are usually relied upon to certify an individual's competence in a certain area.

Career changers may find that with some selected environmental coursework and relevant experience, they can qualify for some types of certification, even without a formal environmental degree. In order to build professional credentials, pursuing certification in an environmental field may be a realistic alternative to earning another formal academic degree.

Following are some examples of some of the current national certification programs and their sponsoring organizations.

- Certified Environmental Professional (CEP) — National Association of Environmental Professionals
- Certified Industrial Hygienist (CIH) — American Board of Industrial Hygiene
- Certified Hazardous Materials Manager (CHMM) — Institute of Hazardous Materials Management
- Registered Environmental Health Scientist/Registered Sanitarian (REHS/RS) — National Environmental Health Association
- Certified Hazardous Waste Specialist (CHWS) — National Environmental Health Association
- Certified Environmental Auditor (CEA) — National Registry of Environmental Professionals
- Registered Environmental Laboratory Technologist (RELT) — National Registry of Environmental Professionals

## EXAMPLES OF SUCCESSFUL CAREER TRANSITIONS

Over the last 16 years, this author has known several individuals who have entered into successful environmental careers from "unconventional" routes. The following list presents some of these successful transitions:

| FROM | TO |
|---|---|
| junior high school teacher | professional geologist |
| Russian studies major | environmental planner |
| housewife | nationally renowned community activist |
| English major | marketer of environmental services |
| medical technician | environmental engineer |
| history major | government environmental program administrator |
| petroleum geologist | environmental hydrogeologist |
| mechanical engineer | environmental engineer |
| news media specialist | environmental trainer |
| state legislator | environmental consultant |

And there are certainly more unusual career transitions than those listed above. The one commonality with each of the individuals involved in these transitions is a personal decision to take the initiative, commit to a goal, and to make it happen.

## ATTITUDE AND MOTIVATION ARE THE KEYS

Before embarking on a professional career change, individuals need to consider that personal sacrifices will be required. Aspiring environmental professionals should carefully assess their interest, motivation, and attitude towards the transition. Implementing the recommendations provided in this chapter will require some, and often substantial, commitments of time and effort. This may mean working nights and weekends. For those whose personal motivation is high and attitude is positive, they should by all means charge on because the opportunities are there!

Once the decision has been made to become an environmental professional, career changers should consider a few additional points.

First, they should be realistic. Setting unrealistic goals will only result in frustration. Second, they need to practice patience. Individuals need to give themselves time to undertake career changing activities, and time to await the results. Third, they should be persistent. Establishing a reachable, specific goal and staying focused on it will provide the best opportunity for success. As the old saying goes, nothing worthwhile ever comes easy. Persistence will eventually pay off, as employers recognize an individual's enthusiasm and interest through repeated contact. Fourth, flexibility is important. Although a specific career goal has been identified, job candidates who are extended offers for other types of environmental work should consider accepting it, if the work can provide valuable experience and build better credentials for the originally targeted career path. Besides, there is a chance that the unsolicited opportunity may prove to be a more interesting or lucrative position. Lastly, and most important, career changers must keep a positive frame of mind. A person who is interested and motivated enough to follow the steps outlined in this chapter definitely has something to contribute to the environmental field. This approach will instill a personal sense of confidence that will make the journey easier.

## CHANGING ENVIRONMENTAL CAREERS

Environmental professionals enjoy the option of relatively free career movement between the public and private sectors, between academia and the "real world" (i.e., government and industry), and among employers in the same sector. This is particularly true for positions which require a specific scientific or engineering background. Because there are few barriers to career movement within the environmental fields, and because of the shortage of available environmental specialists in certain disciplines (translated: more opportunities for higher pay and/or more responsibility), environmental professionals tend to change jobs every three to five years, particularly within the ranks of consultants.

Individuals who stay in one position for more than five years run the risk of being "pigeon-holed" by their current employer or by prospective outside employers. Although the employee may feel he or she is being loyal to the organization by staying in one position for ten years or more, the employer may perceive that the person has only limited knowledge and abilities, which could adversely affect promotional opportunities. Also, in the fast changing and dynamic environmental field, focusing for too long on a single area can result in losing touch with new developments. At the other extreme, professionals who change jobs too often (less than two years), may develop a reputation as a "job-jumper" (who is only looking out for personal gain). This situation will also reduce options for career mobility.

One of the advantages of career mobility is that environmental professionals may be exposed to different types of work and continually find exciting new challenges. A scientist or engineer who has "been around the block" is generally more valuable to the next employer. In larger organizations that house multiple programs, such as state or federal government agencies, environmental professionals have the opportunity to move around among programs every three to five years, without changing employers. Similar opportunities are available with corporations, such as large petroleum, chemical, pulp and paper, and power companies, who staff large environmental affairs departments. If prospective employers see a person's career and job changes as constructively building professional competence, occasionally changing jobs or functional responsibilities within the environmental field can be highly advantageous.

## REFERENCE

Kummler, R., Hughes, C., Witt, C., Stern, B., and Powitz, R., Hazardous Waste Education and Training in the United States, in *Hazardous Materials Control*, vol. 4, no. 3, May/June 1991, p. 16-22, published by Hazardous Materials Control Resources, Inc., 7237 Hanover Parkway, Greenbelt, Maryland 20770.

# CHAPTER 13

# Internship and Volunteer Opportunities

# Internship and Volunteer Opportunities

Employers of environmental professionals are primarily looking for two things when they interview a job candidate, educational preparation and experience. After working at a professional job for a few years, an environmental specialist's experience becomes apparent by the answers to interview questions and information provided on a résumé. But what of the new college graduate or mid-career changer? How do you gain professional experience when those who do the hiring require some experience as a prerequisite? The answer is through internships and volunteering. College students and career changers need to assume that internship and/or volunteer experience is mandatory for employment in the environmental fields.

## INTERNSHIPS

Internships are commonly sponsored by large companies, as well as state and local government agencies. In fact, many colleges and universities arrange for internships with specific government agencies and companies as part of their academic program offerings. Others are arranged through employers, such as manufacturers or consulting firms, who seek out interns for temporary assignments over the summer, or even during the academic year. And finally, some internships are identified only through the initiative, creativity, and persistence of motivated students. They contact employers who perform work in the areas they are interested in, even if the employer never had used interns in the past, and negotiate their own internship arrangements.

An internship, whether paid or unpaid, performed for college credit or not, should provide the aspiring environmental professional with meaningful learning experiences and give him or her exposure to a wide variety of working situations. Students must share the responsibility, along with the employer, of ensuring that internships have redeeming educational value and do not end up being a convenient vehicle for employers to retain "slave labor". As interns, students should seek out opportunities for new experiences and challenges. The successful completion of a well thought out internship program will add a significant credential to a job hunter's résumé.

Participating in an internship is also an excellent way for students to begin building a network of professional and employer contacts. The relationships established during a student's internship program provide that student with an immediate source of information on job openings and advice on how to approach employers about permanent professional positions.

One organization dedicated to the placement of interns into short-term, paid environmental positions is The Environmental Careers Organization, Inc. (formerly the Center for Environmental Intern Programs and The CEIP Fund, Inc.). Since 1972, The Environmental Careers Organization, Inc. has placed over 3,400 students and graduates into intern positions with government agencies, corporations, consulting firms, and nonprofit organizations. From their headquarters in Boston, this organization coordinates program activities through regional offices in Cleveland, Seattle, San Francisco, and Tampa. For more information about internship opportunities with The Environmental Careers Organization, Inc., contact:

The Environmental Careers Organization, Inc.
68 Harrison Avenue
Boston, MA 02111-1919
(617) 426-4375

Another source of information regarding internship opportunities is the National Society for Internships and Experiential Education and their publication, *The National Directory of Internships*, edited by Amy S. Butterworth and Sally A. Migliore. This directory lists internship openings, mostly with universities, nonprofit organizations, education centers, and interest groups, with categorical headings which include: sciences, public interest, international affairs, environment, government, consumer affairs, communications, and business and industry. To order a copy of this publication, contact:

National Society for Internships and Experiential Education
3509 Haworth Drive
Suite 207
Raleigh, NC 27609
(919) 787-3263

## VOLUNTEERING

In areas where internships are not readily available, or where competition for internship positions is stiff, volunteering is another alternative for obtaining environmental work experience. Although federal, state, and local government agencies occasionally offer volunteer opportunities, the nonprofit organizations and environmental interest groups, in particular, welcome volunteers with open arms. Volunteers are used by many of these organizations to handle administrative tasks which may be mundane and routine. Therefore, the challenge for the volunteer is to identify work assignments that will contribute to one's professional development.

Volunteer work can present a mind-boggling array of opportunities. From being an overseer of a nonprofit organization's wildlife refuge, or a hiking trail construction and maintenance volunteer in a national park, to a community survey researcher at a Superfund cleanup site, a data entry volunteer logging data on the location of endangered species habitats, or an administrative assistant helping with mass mailings regarding a critical upcoming vote in Congress, volunteers perform important functions which can have direct or indirect impacts on the environment.

One major positive aspect of environmental volunteer work is that it is readily available in almost every community. The prospective volunteer can pick the organization that focuses on a topic of professional interest and make an inquiry offering to assist with their program. Because there are so many volunteer opportunities available, students can try working in different subject areas with different "employers". This is an excellent way to get exposed to a variety of working situations, and will help individuals in determining personal work preferences.

Another positive aspect of volunteering is that one need not be a recent college graduate to participate. High school students who show interest, enthusiasm, and capability will be able to find volunteer opportunities which will provide them with some early perspectives on environmental work. Career changers may use volunteer opportunities to "test the waters" before making a commitment to "jump". Even retirees will find challenging opportunities to apply their knowledge and skills to "make a difference" through volunteer environmental work. In addition, as is the case with internships, volunteering provides a wonderful opportunity to build a network of professional and employer contacts.

To locate a volunteer opportunity tailored to a particular area of interest, prospective volunteers have many options on places to look. Check out local opportunities with state and municipal agencies, companies, and nonprofit organizations. Regional and national environmental organizations are always looking for volunteers. Many colleges and universities either have such volunteer opportunities, or know of organizations that need volunteers.

For those interested in volunteer opportunities on public lands, the American Hiking Society publishes an annual guide, *Helping Out In The Outdoors*. This directory lists numerous volunteer opportunities with federal and state agencies and nonprofit organizations to assist with campground management, trail maintenance, and ranger duties on public lands across the U.S. For more information, contact:

The American Hiking Society
1015 31st Street N.W.
Washington, D.C. 20007
(703) 385-3252

In summary, internships and volunteering provide a means for students and career changers to develop professional environmental work experience, while having the opportunity to demonstrate personal skills. Although the marketplace for environmental specialists is strong, there is still a substantial degree of competition (in some cases, fierce competition) for the most desirable professional positions. Meaningful work experiences gained through internships and volunteering can provide a competitive advantage for entry-level positions.

# CHAPTER 14

# The Best Prospects for the 1990s . . . The Hot Jobs

# The Best Prospects for the 1990s . . . The Hot Jobs

This book has introduced the reader to a wide variety of employment opportunities available in the environmental field. It has discussed employers, salaries, career changing, and information sources for specific careers. The vitality of the environmental marketplace has also been considered. But, what careers are the best employment prospects for aspiring environmental professionals? What are the really "hot" jobs for the 1990s?

The author has developed a list of the top ten environmental career opportunities for the 1990s. In the career fields highlighted, the demand for environmental professionals clearly exceeds the currently available supply. Although there are no guarantees that individuals preparing for these careers will immediately be employed, it is definitely a seller's market and the opportunities will be there! Here are the author's top ten, presented in no particular order:

1. **ENVIRONMENTAL ENGINEER:**
   Environmental engineers are in demand to analyze environmental problems and develop the conceptual designs for effective solutions, particularly those that are involved in waste treatment, site remediation, and pollution control technologies, such as:

   - industrial and municipal wastewater treatment
   - hazardous waste incineration
   - environmentally safe landfill design
   - treatment systems for the cleaning of contaminated soils and ground water (e.g., bioremediation, vacuum extraction, carbon filtration, vitrification, etc.)
   - isolation of contaminants in the soil or ground water to prevent human or environmental exposure
   - air emission control systems

2. **CHEMICAL ENGINEER:**
Chemical engineers are in strong demand to adapt chemical process expertise to environmental projects, including:

- chemical treatment of hazardous waste
- chemical treatment of industrial wastewater
- identifying or developing safe chemical substitutes for hazardous chemicals used in industrial processes or consumer products
- developing ways to reclaim or recycle hazardous substances within industrial processes
- developing processes which minimize the amount of waste generated

3. **MECHANICAL ENGINEER:**
Mechanical engineers are in demand as key players for the design and construction of pollution prevention or control processes and equipment. Examples of such mechanical engineering applications are

- the design, construction, and installation of air emissions control systems, such as: bag houses to capture dust and particulates, air stripping towers to remove toxic fumes and gases from air discharges, and other technologies
- the detailed design of hazardous waste incinerators, which involves developing conveyance systems to move wastes and ash, fuel burning systems, piping, and other mechanical details
- the designs and specifications for valves, pumps, piping, meters, skimmers, filters, tanks, vents, and other elements of wastewater treatment and sludge management systems

4. **AIR QUALITY SCIENTIST:**
Air quality scientists are in strong demand to help administer and enforce, or to assist industry in complying with, the strict new requirements of the recently passed Clean Air Act Amendments. The specific tasks that air quality scientists are being asked to perform include:

- developing estimates of air emissions using calculations
- designing sampling programs to determine ambient air quality (i.e., the preceding conditions) and to measure the amounts, rates, and concentrations of conventional and toxic air emissions from actual industrial processes
- interpreting analytical laboratory data to determine which pollutants present concern
- simulating the dispersion of air emissions into the environment through the use of computer modeling techniques

- evaluating the cumulative environmental impact of new sources of air emissions
- evaluating and enforcing or preparing air quality permit applications (depending on the employer), and negotiating specific permit conditions for new and existing air emission sources

5. **HYDROGEOLOGIST:**

Well qualified and experienced hydrogeologists are in strong demand primarily to assist with the investigation, characterization, and cleanup of property with contaminated soil or ground water. They are needed to perform the following types of work:

- determining the locations for ground water monitoring wells and soil borings for delineating the extent of contaminants underground
- determining specific soil and ground water sampling methods for a particular site, and geophysical techniques (such as electromagnetics or ground-penetrating radar) to locate buried drums, tanks, and pipelines
- interpreting the analytical laboratory results from soil and ground water samples
- developing, running, and refining mathematical simulation models to predict how contaminants will move through an aquifer

6. **WETLANDS ECOLOGIST:**

Wetlands ecologists are needed to identify, evaluate, protect, and mitigate the adverse impacts upon sensitive wetland resources. They are in demand by government regulatory agencies, environmental groups, landscape architecture and planning firms, and land developers to:

- identify and delineate wetlands where development is restricted through federal, state, or local regulations, and verify that they meet regulatory definitions of a wetland
- assist in finding ways to allow development in the vicinity of a sensitive wetland area without creating any environmentally adverse impacts
- where development in or near wetlands is allowed, if mitigation measures (i.e., creating new wetlands on the same site or area, equal in size and value to ones being adversely affected or destroyed) are implemented, wetlands ecologists conduct research, experiment, design, and test wetland mitigation techniques that will work on a specific site

7. **SAMPLING EQUIPMENT TECHNICIAN**:
Environmental field technicians are in demand to carry out the field aspects of environmental sampling programs and site remediation operations. Examples of this type of work include:

- evaluating various sampling and measurement devices, and selecting the most appropriate ones for specific applications on project sites
- installing, testing, and operating sampling equipment and measurement devices on industrial sites, cleanup sites, in a community, or out in the country (e.g., for measuring water quality in rivers and streams or air quality in forests downwind from a coal-burning power plant)
- taking and transporting samples of air, surface water, ground water, wastewater, sludges, hazardous waste, soil, dust, gases, or any other substance of environmental concern
- performing paperwork tasks associated with field activities, including: recording technical data from instruments, describing sampling protocols, and summarizing field notes

8. **INDUSTRIAL ENVIRONMENTAL MANAGER:**
Environmental managers are in demand by industrial employers to develop, implement, coordinate, and monitor the environmental management programs at industrial plants. Specific functions include:

- conducting or coordinating regulatory audits and inspections of industrial facilities and operations, and taking action to correct deficiencies
- planning and budgeting for waste minimization, waste treatment, and pollution control equipment projects at plant sites
- facilitating of training for plant employees in the areas of pollution control equipment operations, personal protection, hazardous waste handling procedures, and emergency spill response techniques

9. **ENVIRONMENTAL TOXICOLOGIST:**
Environmental toxicologists are in high demand to evaluate the effects of hazardous chemicals on human health and the environment. This is a critical factor in the determination of air emission, wastewater effluent, and soil and ground water cleanup standards. Environmental toxicologists are needed to perform:

- studies and evaluations of the toxic properties of chemicals and environmental contaminants
- determinations of routes of exposure of toxic substances to humans and to the environment
- analyses of concentration levels of toxic chemicals which begin to exhibit toxic affects on living organisms
- determinations of safe levels of exposure to toxic substances to be used in decisions regarding how far to go in cleaning up soil and ground water at contaminated sites and how stringent regulations should be regarding the release of toxins into the environment

10. **INDUSTRIAL HYGIENIST:**

Industrial hygienists are in growing demand as employers become more aware of environmental hazards in the workplace. In their work to minimize worker exposure to hazardous chemicals, industrial hygienists perform the following tasks:

- identify, measure, and evaluate environmental hazards in the workplace that may affect the health and safety of workers
- develop, propose, and implement industrial process controls to reduce or eliminate hazardous conditions, such as fugitive gas or dust emissions from process equipment
- recommend personal protection measures to minimize worker exposures to workplace hazards, such as requirements for wearing chemical-resistant suits, gloves, and face masks when mixing hazardous chemicals

Of these "top ten" careers, only one (wetlands ecologist) is from the natural resources management discipline. This is not to say that there are not opportunities available in the natural resource management professions for dedicated and persistent individuals. However, in relation to the high demand for scientists and engineers in the environmental protection field, the natural resources careers do not provide the same level of opportunity.

Simply preparing for one of these ten careers does not, of course, mean that there will be jobs lined up for candidates before they leave school or their current employer. There is no substitute for personal competence, interest, enthusiasm, and initiative in competing for and winning job offers. However, if a candidate is well prepared and dedicated to professional environmental work in a specific field, chances are good that suitable employment can be found in these ten areas, as well as several of the other career opportunities presented in this book.

The reader will notice that each of the top ten career opportunities identified is a scientific or engineering discipline. This is not a coincidence. Sound technical training provides a solid foundation from which to build exciting, rewarding, and satisfying career paths. Moreover, most entry-level job candidates are evaluated largely on their technical capabilities. However, those who will go the farthest in these careers are those who go beyond their technical training to develop effective management and communication skills. It is this group of individuals who will become the top business and government managers, the chief researchers, and the presidents of consulting firms.

# CHAPTER 15

# Personal Action Plan

# Personal Action Plan

The title of this book promises the reader a practical guide to career opportunities in the 1990s. Each of the earlier chapters discussed the various aspects of employment in the environmental field and surveyed the career opportunities available. If the reader finds a special interest in one or more of the career paths presented, the real value in this guide is in this chapter. Laying out a personal action plan will provide students and career changers a clear, well thought out, road to follow.

**Develop a sound educational background in the environmental sciences or engineering areas —** Even if the chosen career path is not ultimately as a scientist or engineer, a solid background in these subjects provides a foundation from which to make informed decisions on complex, environmental issues. It does not matter whether one is pursuing a non-scientific/non-engineering career in law, teaching, or journalism; environmental science and engineering courses will provide students with some technical insights and an understanding of environmental problems and solutions. This is not to say that every environmental professional must be a technical expert, because that certainly is not so. However, without at least some general understanding of biological, ecological, chemical, physical, geological, and/or engineering principles and applications, the versatility, and sometimes the effectiveness, of environmental professionals is limited. In addition, although the market for environmental specialists is strong, some technical competence must be demonstrated for most entry-level position openings. There will definitely be competition for new positions, and a candidate's technical qualifications are often the most important area of evaluation.

**Develop sound communication (and management) skills —** Many environmental professionals attribute their success to personal communication skills. For technically competent individuals to have maximum influence with decision-makers, and to fully take advantage of their special knowledge and skills, they must be able to write and speak clearly, concisely, accurately, and articulately. Taking college level courses in English, technical and creative writing, public speaking, and interpersonal communications is a good start. However, practicing these skills is critical to professional growth and development. Take or make opportunities to write and speak for an audience. Join a debate team. Volunteer to give presentations in class. Enroll in a public speaking seminar. Write an article for a newsletter, school newspaper, technical journal, or other publication. Practice allows students to improve their abilities, but even more important, it gives them confidence.

**Participate in educational seminars above and beyond routine class work** — In the rapidly evolving environmental arena, a student (or professional for that matter) can never get enough education. Take every opportunity to attend or participate in educational workshops and seminars. These experiences offer exposure to in-depth discussions on current topics of interest from a variety of professional perspectives. There is rarely enough time or resources in college classroom settings to explore specific environmental subjects in detail. For students, registration fees are often offered at reduced rates, or sometimes free of charge. Attendance at educational seminars becomes increasingly important, and more meaningful, for upperclassmen and graduate students, who already have some academic environmental background.

**Get internship or volunteer experience** — As discussed earlier, students should assume that some level of relevant work experience will be mandatory for entry-level environmental positions. With the wide range of opportunities for internships and volunteer experiences, there is no excuse for students not to have environmental work experience on their résumés.

**Become computer literate** — The tremendous volume and complexity of data used in environmental work requires professionals to make efficient use of data processing systems. Although most environmental professionals do not necessarily need to have computer programming or systems analysis skills, they do need, at a minimum, to be familiar with the operation of common office software packages. Developing a capability for using standard word processing, electronic spreadsheet, and data base management software programs will give students a head start in productivity on their first jobs. In addition, software programs that address specific environmental applications (such as air dispersion modeling, 3-dimensional hydrogeologic contour mapping, stream assimilation modeling, and hazardous chemical data base programs) should be familiar to students specializing in those fields. Lastly, students and professionals who find themselves giving frequent presentations often use various graphics software packages to prepare visual aids like overhead transparencies and color slides.

**Know current environmental laws and regulations —** Particularly important for students in the environmental protection and environmental health and safety fields, an overall knowledge of the federal, and relevant state, environmental laws and regulations will enable students to "hit the ground running" in their first jobs (as well as in their employment interviews). Because these are the fields where the most opportunities exist, students, at a minimum, should be familiar with the basic provisions of the Clean Air Act, the Clean Water Act, the Resource Conservation and Recovery Act, the Comprehensive Environmental Response, Liability and Compensation Act (Superfund), the Safe Drinking Water Act, the Emergency Planning and Community Right-To-Know Act, and the Occupational Safety and Health Act (especially as it relates to hazard communication requirements, or Worker Right-To-Know, and personal protection training) and their associated amendments and promulgated regulations.

**Join and participate in the student chapters of professional associations —** Nearly every one of the environmental fields presented in this book has a professional society or association, and most have student chapters and/or discounted memberships for students. By participating in student chapters of these associations, individuals can get some early exposure to their chosen fields of endeavor, as well as a start to establishing a network of professional contacts. In addition, a common activity of professional associations is to publish employment announcements within their newsletters or journals. These job listings, combined with personal contacts with employers, greatly increase the odds for prospective environmental professionals being hired.

**Join and participate in trade associations** — For students who have had enough exposure to the environmental work performed in various industries to determine a personal industrial preference, joining and participating in trade associations may be helpful to develop potential employment contacts. Note, however, that some trade associations do not allow membership by non-company-affiliated individuals. Approaching such groups as an interested student may allow for participation in association activities. Similarly, if one is interested in employment in a specific geographic area, becoming active with the local or state chambers of commerce may yield other opportunities. Join an organization's environmental committee, if they have one. Study the specific types of environmental management issues that the organization's members are dealing with. Get to know people who can help you in finding employment with specific employers.

**Let everyone you know know that you are looking for employment in the environmental field** — This is not an exaggeration. Any one of your personal acquaintances, whether a neighbor, teacher or professor, relative, friend, fellow student, church acquaintance, academic counselor, or fellow alumni, may provide contacts and referrals which can lead to employment. The old adage, "it's not what you know, it's who you know" holds some element of truth for job seekers. In the working world, unfortunately, the best positions do not necessarily go to the most capable and qualified individuals. In fact, many are not even advertised. Do not underestimate the importance of personal contacts.

**Know what is going on in your field of interest** — As students and career changers are finishing up their academic programs or internships, their attention needs to begin shifting from the classroom environment to the working world. At this time, it is important to keep up-to-date on what is going on in one's chosen professional field. Read professional, scientific, and engineering journals to keep tabs on new laws and regulations, technologies, equipment, approaches to problem solving, employers and their business prospects, and other important matters. Having knowledge of current issues, trends, controversies, and other aspects of a professional environmental field can help the job candidate make a desirable impression during an employment interview.

Following these ten steps will greatly increase the opportunities for employment in the environmental field. However, a job candidate's personal qualities can also weigh heavily in the evaluation process. Be enthusiastic, demonstrate a positive attitude, take the initiative to show the employer that you are interested and motivated about environmental work. It's alright to get excited about what you want to do for a living. Use this energy to your advantage; do not hide it behind your résumé.

# APPENDIX A

# EMPLOYMENT NEWSLETTERS FOR CURRENT OPENINGS

*National Employment Review*
735 Providence Highway
Norwood, MA 02062
(800) 638-0014

This is a monthly newspaper type of publication that carries job advertisements, including those for engineers and environmental professionals.

*Environmental Careers Bulletin*
Environmental Careers Bulletin, Inc.
11693 San Vicente Boulevard
Suite 327
Los Angeles, CA 90049
(213) 399-3533
FAX (213) 399-8763

This job listing focuses on employment opportunities in the environmental protection field, primarily with private sector companies.

*Environmental Opportunities*
Sanford Berry, Editor
Box 4957
Arcata, CA 95521
(707) 839-4640
FAX (707) 822-7727

This monthly bulletin lists current openings. The emphasis is on the natural resources professions. It is sponsored by the Environmental Studies Department of the Antioch/New England Graduate School in Keene, New Hampshire.

*The Job Seeker*
Rt. 2
Box 16
Warrens, WI 54666
(608) 378-4290

This bimonthly employment listing details current openings in categories ranging from forestry to environmental science. It also includes available internships.

*Earth Work*
Student Conservation Association, Inc.
P.O. Box 550
Charlestown, NH 03603
(603) 826-4301

This is a monthly magazine incorporating job advertisements with employment-related articles. Subscriptions are $29.95 per year.

# PERIODICALS CONTAINING ENVIRONMENTAL EMPLOYMENT LISTINGS

*BUZZWORM, The Environmental Journal*
P.O. Box 6853
Syracuse, NY 13217-7930
(303) 442-1969

The "Connections" section of this magazine includes listings of volunteer opportunities and environmental jobs.

*Engineering News-Record*
McGraw-Hill, Inc.
1221 Avenue of the Americas
New York, NY 10020
(212) 512-2000

This weekly engineering news magazine regularly carries employment advertisements for engineering positions. Each November, it carries a special section on environmental career opportunities.

Note: Nearly all of the professional association periodicals, such as the *Journal of the Air and Waste Management Association,* and environmental trade magazines, such as *Pollution Engineering* and *HAZMAT World,* regularly carry job advertisements for environmental professionals.

# APPENDIX B

## OTHER REFERENCES FOR ENVIRONMENTAL CAREERS

*Occupational Outlook Handbook*
U.S. Department of Labor
Bureau of Labor Statistics
600 E Street N.W.
Washington, D.C. 20212
(202) 272-5381

This is the U.S. Government's reference book on careers and employment.

*Outlook 2000*
U.S. Department of Labor
Bureau of Labor Statistics
600 E Street N.W.
Washington, D.C. 20212
(202) 272-5381

This document analyzes national trends in employment.

*Opportunities in Environmental Careers*
by Odum Fanning
VGM Career Horizons
NTC Publishing Group
4255 West Touhy Avenue
Lincolnwood, IL 60626-1975

This book, written by a freelance science writer, includes discussions of environmental philosophy, trends, and innovative college programs.

*The Complete Guide To Environmental Careers*
The CEIP Fund, Inc.
Island Press
Suite 300
1718 Connecticut Avenue N.W.
Washington, D.C. 20009

Along with general career guidance discussions, this book relies heavily on interviews with dozens of environmental professionals to provide a picture of employment opportunities.

*Becoming An Environmental Professional 1990*
The CEIP Fund, Inc.
68 Harrison Avenue
Boston, MA 02111
(617) 426-4375

This book is a compilation of articles of employment trends and issues in the environmental field.

*1991 Conservation Directory*
National Wildlife Federation
1400 Sixteenth Street N.W.
Washington, D.C. 20036-2266
(202) 797-6800

This is an excellent resource book for job hunters. It lists, and provides addresses for, federal and state government agencies, nonprofit associations, and international organizations.

Chronicle Guidance Publications, Inc. publishes a series of *Chronicle Occupational Briefs* for over 600 job titles, including dozens of environmental occupations. Each brief is reviewed and edited by several authorities, and is approximately four pages in length. To request a copy of their current listing of briefs, call their toll-free service line at 1-800-622-7284, or write to: Chronicle Guidance Publications, Inc., P.O. Box 1190, Moravia, New York 13118-1190.

# Index

# Index